Maths and Humour

Edition Angewandte
Book Series of the University of Applied Arts Vienna
Edited by
Gerald Bast, Rector

edıtıon:'ʌngewʌndtə

Universität für angewandte Kunst Wien
University of Applied Arts Vienna

Georg Glaeser & Markus Roskar

Maths and Humour
Solving everyday problems with mathematics

DE GRUYTER

Table of Contents

**Can mathematics be funny –
or, can it at least be fun?**

Mathematics with humour

Mathematics has gained a unique reputation. Mathematicians are widely considered to be intelligent, but just as common are accusations of their being eccentric, insisting on meticulous precision, and sitting in an ivory tower. Such accusations are usually followed by the following statement: "I really sucked at mathematics in school", or alternatively: "I was actually pretty good at calculus".

While mathematics should by no means be reduced to calculus, it is just as wrong to think that it exclusively consists of proofs or considerations that an amateur will find impossible to grasp. Mathematics should not be regarded as being limited to the ivory tower of academics, secluded from everyday reality – it's certainly not forbidden to enjoy mathematics and even have fun with it. Neither is it against the rules to discuss mathematical problems in a manner that the average person will not find difficult to understand.

This book is aimed at people with a personal interest in mathematics, regardless of how many years of mathematics they have completed at school. Teenagers with a talent for mathematics who enjoy spending their free time with a sheet of paper and pencil solving mathematical problems, rather than pursuing other kinds of leisure activities, will probably find this little book particularly interesting. A little piece of advice: sketches are always quite useful in mathematics. You can also have some fun by adding humorous elements to your sketch!

The texts were written by Georg Glaeser, who teaches mathematics at an art university and is usually successful in turning his classes into a delightful experience for his students. Mathematics and the arts are by no means as diametrically opposed to each other as they were long believed to be, after the demise of the universalists (Leonardo da Vinci).

The drawings were made by a colleague of Glaeser, Markus Roskar. It is always worth having a closer look at these drawings: they all contain some punch line or hidden joke. So, even if you find a text in this book lacking in humour, the drawing will certainly make up for it by putting a smile on your face. The two authors wish to express their special thanks to Tamara Radak, an invaluable advisor for this publication in all its phases, and her co-translator and co-(copy-)editor Eugenie Theuer. Other contributors who should be mentioned here by name are Peter Calvache, Max Gschwandtner, Boris Odehnal, and Günter Wallner.

The design of this book is based on the double-page principle: each double page consists of a drawing, which is usually quite humorous and relates in some way to the text on the opposite page. The text is meant to be informal, often amusing while still mathematically correct. It will often relate to some issue of everyday life that has a mathematical dimension. This design offers several advantages: Such a double page can well be digested before breakfast or be enjoyed as a bedtime story that will leave a smile on your face as you drift off to sleep. The double pages have been designed to be self-contained so that they can be read independently and in any order. So, take this as your cue and have fun leafing through this book!

The oracle's instructions
and their implementation

ΑΡΧΥΤΑΣ

ΠΛΑΤΩΝ

Making of

Mathematics is currently shedding its image as an inaccessible, cumbersome science and acquiring an open and emotional dimension. So, the time has come for us – the artist Markus Roskar and the maths professor Georg Glaeser – to write a "maths book of a different kind".

As lecturers at an art university, we are used to having a very diverse audience: what matters at the entrance exam are not so much the grades that our students receive in mathematics, but their artistic creativity. So, getting to know the students selected in this process and finding out about their relation to mathematics is always quite interesting for us. Most of the students are young people who are actually quite interested in applying mathematical principles without being suffocated by a tight corset of strict terminology and definitions. They enjoy humorous classes about mathematical tools that are suitable for everyday use and also like being surprised by unconventional examples.

In order to write a classical mathematical book, you carefully consider the basic framework for your book, divide it into chapters, and then think about the order in which these chapters are presented to guide the reader step by step from one level of your argument to the next. Even if you only want to cover a specific area of mathematics, your book will have at least a few hundred pages and it will take the reader a lot of time to read. There are, of course, plenty of such books on the market, but they are unlikely to be read by the audience that we are targeting. It has been our aim to create a handy book that you can carry around with you and open to a random page to get some inspiration.

This has not been an easy process, as the pile of drawings lying on Markus Roskar's desk attests. Only a fraction of these drawings have "made it into the book" – and so did only some of Georg Glaeser's theoretical contributions. Sometimes the originality of a drawing became a decisive factor and set the tone for the text. At other times, it was the mathematical message of the text that was given priority over some funny detail in the drawing.

The series of drawings on the left page would not have made it into the book if we had not decided to include a "Making of" double page. The drawing deals with a classical problem of the ancient Greeks, namely the doubling of the cube, also known as the Delian problem: during a plague epidemic, the citizens of Delos asked their famous oracle for advice. The oracle then told them to double the volume of the cubical altar inside the temple of Apollo. Nowadays we would just type the cube root of 2 into our calculators but in ancient times the problem was demonstrably impossible to solve with classical methods – that is to say, it could not be constructed with ruler and compass.

1
Elementary

Mathematics and other sciences

Mathematics was not developed solely for its own purposes (today it is considered to be the most developed field of study). It was always supposed to contribute to the development of other sciences. In fact, the triumph of mathematics extends even into the furthest corner of many other sciences, including biology, geography, medicine, and musicology.

Quite naturally, mathematics is firmly rooted in the field of physics. Physicists are sometimes called "semi-mathematicians" – though one should bear in mind that there is not only theoretical physics, but also practical physics. So, it is perhaps no surprise that there are many mathematical jokes involving physicists.

Here's a good example of such a joke: three mathematicians and three physicists are travelling to a conference by train. To the mathematicians' surprise, the physicists have bought only one ticket for all three of them. When they catch a glimpse of the guard, the three mathematicians disappear into the wagon's toilet. The guard notices that the toilet is locked; so, he knocks on the door and asks for the ticket. When the physicists slide the ticket through the crack of the door, the guard validates the ticket before sliding it back and then moves on to check the tickets of the other passengers.
After the conference, the two groups of three are returning home by train. The three physicists have once again bought only one ticket, but this time the mathematicians have bought *no ticket*. When they see the silhouette of the guard approaching their wagon, the physicists lock themselves into the toilet again. One of the mathematicians then walks to the toilet door, knocks, and says: "Tickets, please!"

In our list of sciences and scientists that draw on mathematics, we almost forgot those that are most indebted to this discipline, namely engineers. Engineers love mathematics – for a good reason – and they continually apply it when building bridges, optimizing machines, calculating electrical circuits, etc. However, from a mathematician's perspective, engineers are often not precise enough:
Once again a group travels by train, but this time they are on their way to Scotland: it consists of an engineer, a philosopher, and a mathematician. Through the window, they spot a black sheep in the midst of a white flock. The engineer quips that Scotland also seems to have its black sheep. The philosopher admonishes him for making such a generalization, telling him to be more precise and say that they can only speak of *at least one black sheep*. The mathematician, however, cannot accept this statement either and rephrases it by saying that there is *at least one sheep* that is coloured black on *at least one side*.

The abovementioned jokes usually elicit a laugh even if the audience members do not belong to either of these occupational groups. Some jokes that can be found online, however, do require some specialist knowledge usually exhibited only by maths freaks.

Not the same but certainly not a contradiction ...

Mathematics and art?

This question is regularly raised *outside* of universities for art and design. It may be quite clear for anyone that geometry comes in handy for industrial designers and architects. But what about the other, more abstract fields of mathematics?

Many an artistically gifted person (and also others) show a certain reluctance toward mathematics, which often goes back to unpleasant memories associated with the subject from their school days. During the Antiquity and the Renaissance, geometry and mathematics were by no means viewed as being diametrically opposed to the arts, as evidenced, for instance, by the work of Leonardo da Vinci. In recent years, chances for the convergence of mathematics and art have improved as institutions are increasingly becoming aware of the benefits of such a symbiosis. Ideally, the result of this convergence would be a mutually enriching interaction and development of the two disciplines, occasionally even the breaking of established taboos. We could call this process an evolution based on reciprocity, and it is a trend that can be observed now well beyond the confines of art universities.

The impact of mathematics on the arts is by no means restricted to the creation of works expressing artists' fascination with mathematical formulas and surfaces, whose aesthetic value is indisputable. Sometimes it is the astonishing precision underlying all mathematical statements that serves as an inspiration to artists.

There is a famous quote by the American poet and writer Charles Bukowski which says: "The problem with the world is that the intelligent people are full of doubts while the stupid ones are full of confidence." With a note of humour, let us consider the following mathematical questions in light of Bukowski's statement:

True artists are always full of doubts. Does it follow from this that artists are always intelligent? Or would a correct conclusion be that people who are always sure about themselves and their own opinions must, therefore, be stupid? (Mathematicians sometimes tend to view things as cast in stone, even if these things have not originated in the artificially developed construct that is mathematics.)

These questions are not easy to answer, but they are quite thought-provoking, which is an effect that art frequently has on mathematicians and vice versa. Bukowski's quote may remind German readers of a proverb in their language that can be translated as: "Stupidity and pride grow on the same tree." The problem with proverbs is that they often lend themselves to different interpretations. Yet, there always seems to be a (fairly large) grain of truth in them.

Can there be a more perfect shape?

Mathematics, art, and beauty

You will already have noticed that this book is not a classical math book. Neither is it a book about art or beauty. Nonetheless, these three terms – math, art, and beauty – keep colliding with each other here, and mathematicians do, of course, see a certain beauty in their formulas and figures. Yet, they would rarely refer to such things as art. It is difficult to define what art is. The oft-repeated German saying "Kunst kommt von Können", which can be translated as "Art comes from artisanship", is usually used to ridicule an artist. However, the phrase that is occasionally quoted by politicians to counter that saying, "If art comes from artisanship, then Carly comes from Caorle", is not completely correct, either: the German word "Kunst" (art) is, of course, etymologically related to both "Kennen" (knowing) and "Können" (skill, artisanship). The following observation was recently made by an artist: "If somebody is an artist, they must know everything." Another skilful play on the German words "Künstler" (artist) and "Kennen" (knowing).

When designers and architects wish to incorporate something similar to an ellipsoid in their designs or drafts for a building, the common advice "Take an exact ellipsoid, because nothing really beats its aesthetic" is actually quite valid. "Rough approximations" are often the only flaws in a building. Another no-go that is often mentioned in architecture or design magazines are vertical edges that are "not quite parallel". It is better to go the whole hog.

There are also some interesting opinions on symmetry. A new mathematical formula with symmetric coefficients makes the heart of every mathematician or physicist beat faster. This preference for symmetry is often taken to such an extreme that you can hear mathematicians say, "If a formula is not symmetrical, there is ground to suspect that it might be incorrect." Sometimes these people are actually right, and you can find some minor errors in the derivation of the formula.

Even in supposedly airtight computer graphics of mathematical figures, you can occasionally find errors. When surfaces in space described by mathematical equations are considered locally, they sometimes deviate from the expected shape and seem quite odd. The simplest explanation for this is an error while typing the equation. The image to the left shows a so-called trinoid. Those who have tried to type the equation for such a shape are probably quite familiar with the issue of typing errors.

Such inconsistencies might also shed light on the shortcomings of computers: computers can usually calculate numbers only up to a certain number of decimal places. So, for instance, if you try to subtract two numbers that only differ from each other from the tenth decimal place onwards but have been half-adjusted by a computer, you are likely to end up with some "digital gibberish".

The ``instruments´´ are
in tune: let's go!

Mathematics and music

A t all times and in all cultures, humans have been known to break into song and make music. Indeed, some people seem to have an inborn talent for rhythm and they deliver sounds and sound sequences that go straight to the heart. (There are, however, exceptions to the rule, such as the Gallic singer Troubadix, known from the comic series *Asterix*.) These people are often not capable of reading music at all, simply playing or singing by ear.

A sound is produced when air oscillates in such a way that the ensuing sound waves have a certain frequency. That can occur due to the oscillation of a tensioned string or, e.g., by blowing on a bottle that is partly filled with liquid (see the drawing on the left). The famous "concert pitch" has a standard voice level of 440 Hz. That's not entirely true – a number of famous orchestras have a standard pitch of 443 Hz, which lends a richer sound to string instruments. Though the difference is marginal (a tenth of a halftone), people with absolute pitch are able to recognise it.

T he fact that Pythagoras already divided sounding, tensioned strings according to simple mathematical proportions (1 : 2, 2 : 3) 2,500 years ago, thus defining a scale, demonstrates just how closely music and mathematics are interlinked. Now let us consider those frequencies that are indirectly proportional to string length. A frequency ratio of 2 : 1 equals half the string length and the jump from the lower to the higher sound is called an octave (e.g. C$\mapsto$ c). When we are dealing with a fifth, the frequency ratio is 3 : 2 (C$\mapsto$ G) and for the fourth, it is 4 : 3 (C$\mapsto$F or E$\mapsto$A).

M usical scales and tonal systems have been adapted in the course of the history of music. A pure-Pythorean temperament sounds a bit odd to our ears today. This is related to the fact that simple fractions of integers are not a perfect fit for the exponential scale of the frequencies. That is reminiscent of Fibonacci numbers (see p. 117), which provide a good, yet not perfect approximation of exponential growth in nature. An exact mathematical division is achieved in dodecaphony (usually a bit of an acquired taste).

M athematicians like to claim that a disproportionate number of their "guild" display extraordinary musical talent. The authors of this book were not able to verify whether there are serious statistics supporting this assumption, which is complicated by the fact that musical talent is very much a subjective matter. It is similarly difficult to define a talent for mathematics. Both of these talents lie dormant in most people, and both can either be fostered or alternatively fade to the background.

(Mumbling:) And yet it moves!

Being adamant about something

Italy used to be a hotspot for geniuses during the Renaissance. Leonardo da Vinci was one of them, Galileo Galilei another. Their places of birth, Vinci and Pisa, are situated at a mere linear distance of 40 km; however, Leonardo was born more than a century before Galilei. Both of them were universalists – or, indeed, Renaissance men in more than one sense of the term –, going far beyond the state of play in the sciences at the time. Both of them sailed close to the wind with their ground-breaking theories.

In their time, it was quite easy to come into conflict with the powerful clergy when looking for explanations for occurrences that did not quite make sense. Leonardo, for instance, used to dissect corpses in order to better understand the relationship between individual parts of the human body. Galileo, on the other hand, took up Copernicus' century-and-a-half-old theory, which did not position the earth but the sun in the centre of the universe and explained the observed inconsistencies in planetary orbits. Copernicus' work had been derided at first – however, it was later put on the Index of Forbidden Books (*Index Librorum Prohibitorum*) for good measure (apparently as a result of the recent "troubles" with Galilei) and remained there for more than two centuries. Copernicus was probably unaware that Aristarchus of Samos had already propagated the heliocentric system 1,800 years earlier, his idea similarly falling on deaf ears.

Of course, there's a difference between being ridiculed because a theory is difficult to accept and being burned at the stake for it. Galilei was pragmatic enough to backtrack on his views at one point. Perhaps he sensed that the hypothesis that the earth may be the centre of the universe could not hold much longer anyway. Soon, Johannes Kepler published his famous laws of planetary motion, which still later led to Isaac Newton's law of universal gravitation.

What about today, then? In some states of the US, the Theory of Evolution appears in school books only as one of a number of theories explaining the diversity of species – on a par with untenable hypotheses that practically question everything that can be considered the scientific state of the art.

Pure mathematicians, who are not concerned with physics or biology, however, never had any problems relating to intolerance or prohibitions. Their formulas do not threaten conservative world views, as they are divorced from reality in some sense. Their formulas are neutral because mathematics is a self-contained microcosm that does not necessarily bear ad-hoc connections to reality.

``The first 100 lines of the proof are okay. The last ones, too. Now we just need a miracle for the middle part ...''

Every line of proof must be airtight

Mathematical jokes circulate – like many other kinds of jokes – on the Internet. While mathematicians are usually quite amused by them, other people often manage no more than a weak and forced smile. On the left-hand side, we see a drawing of a group of mathematicians standing in front of a blackboard scrawled over with a mathematical proof. The first and final lines of the proof work seamlessly, but in between, the proof still needs a "missing link" to be completed. While non-mathematicians may not consider this to be a joke, mathematicians will be painfully familiar with this kind of situation; with a smile, they might reminisce about their own attempts at figuring out proofs in which "almost everything" seemed to work. But just one faulty line is enough to upset the entire calculation.

Here are two simple examples that are often cited as classics of such faulty proofs:

• You could use the following two lines to prove that 1 equals 2 – if it wasn't for one mistake ...
$$a = b \Rightarrow a + (a - 2b) = b + (a - 2b)$$
$$\Rightarrow 2(a - b) = 1(a - b) \Rightarrow 2 = 1.$$
With a solid education in mathematics, it should not take long to find the mistake, but an amateur might struggle a bit: in the final step, we divide by $(a - b)$, which is only allowed if $a - b \neq 0$ and thus $a \neq b$ (as is well known, division by zero is not permitted because its result is indeterminate). Yet, we started with the premise that $a = b$ ...

• Even an amateur can easily find the mistake in the following example of "how to waste money":
$$1€ = 100\,c = 10\,c \cdot 10\,c =$$
$$= 0.1€ \cdot 0.1€ = 0.01€ = 1\,c.$$
The mistake here lies in the so-called dimension: rather than $10\,c \cdot 10\,c$, we should have written $10 \cdot 10\,c$ ...

Many people believe that it just takes perseverance to arrive at a solution to a given problem, even if there is mathematical proof that there can be no such solution. Let me give you a simple example: no set of three odd numbers from the numbers $1, 3, 5, 7, ..., 93, 95, 97, 99$ can add up to 100. Since all the numbers are odd and the sum of odd numbers is always odd as well, the result can never be an even number.

What may seem obvious here can sometimes be much more challenging to figure out. Time and again, people claim that they have found a way of determining the exact value of π by means of ruler and compass only. There is a mathematical proof to refute such claims, but it is difficult to understand. To show that a geometric calculation of π is not 100% exact, it is often enough to repeat the same construction process on a computer and then have the computer calculate the value to 10 decimal places. Many people put more trust in the calculation capabilities of computers than mathematicians. A result that is correct up to 5 or 6 decimal places would already be quite remarkable.

2

That's Genius!

Albert is doing handicrafts.

Does a single proof suffice?

The ancient Egyptians were well familiar with the theorem $a^2 + b^2 = c^2$, which defines the relationship between the sides of a right triangle. They knew the theorem, but it did not occur to them that they might have to prove it. It was only 2.500 years ago that Pythagoras, after whom the theorem is named, developed a proof (in recent years, the single-handedness of the Greek mathematician's achievement has been called into question). The ancient Egyptians would often apply "theorems" without noticing that these held only approximately.

By now, dozens, if not hundreds of different proofs of the Pythagorean theorem have been put forward. It is always good to have several independent proofs of the same theorem, although a single proof actually suffices. However, it is worth having a closer look at the proof developed by then 11-year-old Albert Einstein, because it is quite remarkable. Young Albert was a mathematically and geometrically gifted child, though there have been false claims to the contrary based on a misinterpretation of Einstein's matriculation certificate. His first biographer did not realise that 6 is actually the *highest* possible grade in Switzerland!

The boy was clever enough to understand the following: when a plane figure is scaled (= multiplied) by a certain scaling factor, the *surface* of that figure is enlarged by the *square* of the factor. For example: a circle of radius 3 has a surface area that is nine times as large as that of a "unit circle" of radius 1. No formula for the surface area is needed here ...

Here is another (no longer trivial) example: we want to fold a sheet of paper in the middle to get two similar sheets with half the surface area. The formats are similar to each other (same aspect ratios). So, when scaling up from A5 to A4, the side lengths increase by the square root of 2, as $(\sqrt{2})^2 = 2$.

Young Albert thus came up with the following idea: a right triangle with the sides a, b and c is given. By cutting this triangle with a pair of scissors along the height of its longest side c, it can be divided into two smaller right triangles. When these two triangles are combined, their surface area is the same as that of the original triangle. The triangles even have the same angles and are thus *similar*. They are also similar to a further "prototype" triangle with the same angles. The longest side of this prototype is assumed to be 1 and its surface area is F. So, by scaling the longest side of the prototype by the factors a, b, and c, we will get our three triangles from the beginning, and the surface areas of these triangles must, therefore, be $a^2 \cdot F$, $b^2 \cdot F$ und $c^2 \cdot F$. Since the original triangle is made up of the two smaller triangles, the following holds true: $a^2 \cdot F + b^2 \cdot F = c^2 \cdot F$. Now we only need to divide by F in order to complete our proof!

When every single step of a proof is accepted as reasonable by a group of mathematical experts, chances are that the theorem to be proved will hold for eternity. That distinguishes mathematics from almost any other science.

$10^3 + 9^3 = 1729 = 12^3 + 1^3$

How can you think that 1729 is a meaningless number?

A knack for numbers

Srinivasa Ramanujan was born in India in poor circumstances and died at 32 at the same place, having become a mathematical celebrity in England but being ailing throughout his life. His handling of numbers was legendary. A mathematician friend of his is said to have visited him in hospital and mentioned in passing that he had arrived in a cab bearing the number 1729 – and that this number was entirely meaningless to him. Not missing a beat, Ramanujan is supposed to have replied: "No, the number 1729 is quite remarkable – it is the smallest natural number that can be expressed as the sum of two cubic numbers in two ways!" Here is an illustration of this statement for laypersons:

$$1729 = 1^3 + 12^3 = 9^3 + 10^3.$$

Not bad, right? There's more: the following formula by Ramanujan makes it possible to calculate π down to the 88th decimal place in just ten steps:

$$\frac{1}{\pi} = \frac{2\sqrt{2}}{9801} \cdot \sum_{n=0}^{\infty} \frac{(4n)!}{(n!)^4} \cdot \frac{1103 + 26390\,n}{396^{4n}}.$$

There is no need to understand the formula as such – it was and still is a brain-teaser even for mathematicians. And yet, Ramanujan was an autodidact, gradually having to learn just how rigorously mathematicians approach these issues. The man was evidently a natural, these things falling into his lap.

His notebooks are brimming with such formulas, many of them without proof.

At this point, a mathematician may think of *Fermat's Last Theorem*: it deals with *integer* triples (a, b, c) complying with the equation $a^n + b^n = c^n$. In the seventeenth century, the brilliant French mathematician Pierre de Fermat claimed in one of his notebooks to have found an "exquisite proof" demonstrating that there is no such triple for $n \geq 3$. He merely noted that the proof itself was quite long and would not fit on the page …

A short remark: for $n = 2$ we have Pythagoras' theorem $a^2 + b^2 = c^2$ in a right triangle. The ancient Egyptians were already familiar with the integer triple $((3, 4, 5)$: $3^2 + 4^2 = 5^2$. With this, they were able to measure their rice fields each year after the Nile floods highly efficiently. They used ropes of length $3 + 4 + 5 = 12$, tied together the ends and stretched it between three people at the initial point, after 3 knots, and after another 4 knots. The ancient Babylonians produced the integer triple $(5, 12, 13)$.

The definitive proof of Fermat's Last Theorem $(n \geq 3)$, a seemingly simple formula, brought many a mathematician to the brink of desperation. It was not until 1994 that the British mathematician Andrew Wiles finally succeeded in providing a proof.

Witch or wunderkind?

Child prodigies

Maria Gaetana Agnesi was born almost contemporaneously with Austrian Habsburg ruler Maria Theresa (whose dominion extended to Milan) and therefore during the Enlightenment period. She was lucky in that respect, as had she been born one or two centuries before, a mistranslation could have led to her quick demise at the stake: Among other things, she discussed an algebraic curve that she called "versiera", which translates as "witch" in colloquial Italian (the correct term would have been *versoria*). The curve is still known as "witch of Agnesi" in English. Maria Gaetana was the eldest of 21 (!) siblings and a so-called child prodigy: by the age of 11, she was fluent in seven languages and showed an extraordinary talent for mathematics. She was appointed as a university professor at the University of Bologna by the age of 30.

It was a novelty for women to be offered such positions and it took until the late 20th century for women to be taken seriously in this field, as in other sciences. The reasons for this would go beyond the scope of this book and would not exactly fit the topic of humour either ...

Agnesi, however, never actually practiced as a professor. She abandoned her scientific career in favour of her faith and her charitable work.

Agnesi's life story is partly reminiscent of Blaise Pascal's – he lived a century before her. He, too, was a child prodigy, advancing the field of mathematics at the tender age of 16 with an impressive publication on conics. Pascal was sickly from an early age. He interpreted his ailment as a sign of God and began leading an ascetic life, in which maths played a minor part compared to religion. He died at the early age of 37.

There is a well-known anecdote surrounding the young Carl Friedrich Gauss. One day, he was given a task as punishment: He was supposed to add up all numbers from 1 to 100. By writing down (or rather outlining) the numbers 1 to 50 from left to right and then the remaining numbers from 51 to 100 from right to left, he realized that any two numbers lying above and below each other yielded the sum 101. The result was thus $50 \cdot 101$. The corresponding formula is still known as "Little Gauß" in the German-speaking world.

Gauss serves as an example that mathematically talented child prodigies do not "inevitably" end up sacrificing their penchant for this science over their life course. Indeed, he became one of the greatest mathematicians of all time. Gauss is also the originator of the phrase "Nothing shows a lack of mathematical education better than an overly exact calculation."

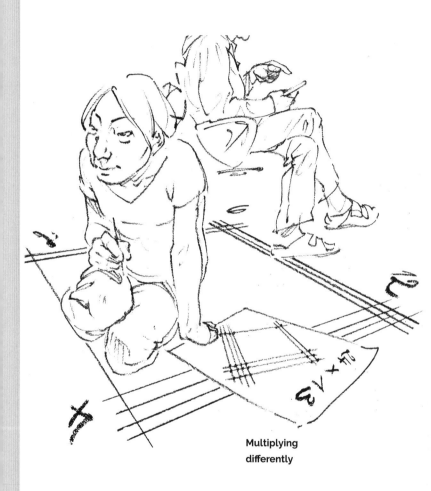

**Multiplying
differently**

How was multiplication done in the olden days?

Today, we are used to multiplying numbers with an app on our mobile phone. We used to do it with a calculator, and, still before that, using calculating machines that had to be cranked. Or simply with the help of pen and paper, obeying the rules set up by Adam Ries. His book was among the first ever printed matter, and it was published in more than a hundred editions until the 7th century. Be honest – are you still able to do multiplication manually? Can you get the correct result immediately? Don't worry – this has very little to do with intelligence; it is basically a recipe, or, mathematically speaking, an "algorithm". It is unlikely that the polymath Abu Dscha'far Muhammad ibn Musa al-Chwārizmī, who taught at the "House of Wisdom" in the ninth century, could have predicted that "mathematical recipes" would one day be named after him ...

Our numeral system has been based on the decade system since the Arabic times, which revolves around the invention of zero. The Arabs adopted the number zero from the Indians, who had been calculating with it centuries before. This is how the invention eventually arrived in Europe in the twelfth century. The invention was as genius as it was simple: attaching a zero at the end of a number equals a tenfold increase in its value. In the process, the digit that came before slides to the left. The position of the digit was thus crucial, which means that ten digits were enough: $0, 1, 2, \cdots, 9$ (see also p. 131). For these, efficient rules were devised with which it was possible to carry out the four basic arithmetical operations.

This leads us back to Adam Ries.

These considerations raise a burning question: how did, for example, the ancient Romans do mathematics? With the plump system of Roman numerals, things moved at a sluggish pace. An example: the last female pharaoh, Cleopatra VII, was born in the year 684 a.u.c. (*ab urbe condita* = from the founding of the City of Rome 753 BC). The Romans calculated her age at the time of death with the help of her date of death, 723 AUC: $DCCXXIII - DCLXXXIV = XXXIX$. Mathematicians could, of course, also find an algorithm for this, but it would be so complicated that it would not be particularly practicable.

The answer to the problem was the abacus – a counting frame (the Romans in fact invented a "mobile" abacus, which could be hidden under one's toga). By moving around the beads, it is indeed possible to carry out the four basic arithmetical operations. Its method of operation is quite complicated and not easily explicable in a few lines. The abacus is, however, surprisingly efficient and used in many cultures, e.g. in Japan.

Speaking of Japan: in the picture on the left, a young woman is demonstrating the multiplication of two numbers. The tens and the unit positions are arranged in parallel at intervals, then all intersection points are counted and their numbers are summarized on diagonals. Some carryovers may be inevitable ...

Can you please leave me alone?
After all, in 140 years, a computer language
is supposed to be named after me!

A 200-year old computer programme?!

Ada Lovelace – actually, Augusta Ada King, Countess of Lovelace (née Byron) – lived in the first half of the 19th century in London and was highly gifted in the field of mathematics. Owing to cancer, she did not even live to see her 37th birthday. Not ever meeting her father, the poet Lord Byron, Ada was brought up by her mathematically interested mother and was homeschooled by her teachers in the natural sciences and mathematics as well in her youth. At 17, she paid a visit to the mathematical salon of the mathematician and inventor Charles Babbage, who is the reason why Ada is present in this chapter.

Babbage, who became Lovelace's husband later on, facilitated her access to mathematical libraries (women were barred from libraries at the time) by transcribing articles. Nowadays, in the age of the Internet as well as a range of copying machines, this is difficult to imagine. Despite this support, it became increasingly difficult for Ada Lovelace to balance time spent with her three children with her passion for mathematics and music – this will sound quite familiar to some of our female readers. During the final years of her life, Lovelace is said to have done away with social conventions and, among other things, to have dabbled in horserace betting, developing a mathematically sophisticated and "reliable" betting system (see also p. 135).

But let us go back to the cryptical reference in the context of Charles Babbage. He developed an "analytical machine" but did not live to see it constructed. The underlying idea was a calculation of tables for use in the natural sciences and engineering. The Italian mathematician Luigi Menabrea was quite taken with Babbage's idea, and he wrote an article about this (at that time still hypothetical) machine in French. Ada Lovelace was supposed to translate this article; however, she surpassed herself during this task, expanding it to twice its original size by means of her own comments and developments. In this context, Lovelace developed what could be termed the first simple formal computer programme: a written plan for calculating the Bernoulli numbers in the form of a diagramme (by the way – the numbers are related to the Riemann zeta function; see p. 73).

The young lady realised that with the use of programming cards, the analytical machine would surpass the limits of simple calculating machines by far. Ada recognised the far bigger potential of the machine: it would not only be capable of conducting numerical calculations but also of processing letters and producing music. She already differentiated between hardware (the physial part of the machine) and software (the coding of the punchcard).
Pretty radical for her time, right?

Everything's on track!

Simple yet ingenious

What keeps trains from derailing all the time? What a stupid question – they move on rail tracks, of course! After all, we often use the phrase "Everything is on track." But there is also an ingenious mathematical (geometrical) explanation for this. Or, as they say: simple yet ingenious!

Let us imagine an asymmetrical dumbbell rolling across the floor. Its two "wheels" have different radii (let the right wheel, for instance, be the larger one). If we now push the dumbbell on the right-hand side, it will automatically make a rotation towards the left. Could we use this observation to build a dumbbell with an "automatic correction"? This dumbbell would initially have two wheels of equal size. If the dumbbell tilts towards the right, we will promptly enlarge its right wheel, which will cause the dumbbell to swerve to the left and get back on track again. Its two wheels can now return to their original, equal size. Hightech?

The practical implementation of this thought experiment does not require any complicated technology as long as there are rail tracks and we apply the following trick: if the outside rim of a train's wheel has a *conical shape*, as depicted in the drawing on the left, the radii of the contact circles between the cone and the rails could vary and the "dumbbell trick" would still work. A slight tilting of the train to the right will cause the radius of the right contact circle to become larger and the radius of the left circle to become smaller. This will correct the train's movement and push it back to the left. It works the same in the opposite direction and the train is thus forced to "remain on track" without the aid of a computerised control system.

The wheel's steel tire is only meant to serve as a safety measure for sudden emergencies such as a forceful blow from the side or when the train goes too fast while taking a turn and the curve is not sufficiently inclined.

Speaking of which: Does the same principle enable a train to take a turn? If it is a curve of constant radius, only the ratio of the contact circle's radii has to remain constant in order to keep the train from derailing. That is why train wheels are quite unsymmetrical while moving on curved tracks. If the train goes at the right speed, this will ensure a stable position. However, if the train goes too fast or too slow, it can be thrown slightly off track due to the centrifugal or gravitational forces at play. Yet, unless it deviates from the right speed too drastically, the train's "autocorrect system" will push it into the right position.

Unfortunately, this system has already been established as a standard – otherwise, we could have made a fortune with this clever invention …

3
Rough Estimates

Ten times to the moon and back...

Eight billion

This is probably one of the double pages where our attempts at humour may occasionally falter. However, you will find the accompanying drawing quite original, and recent years have at least shown first signs of a trend reversal ...

At the beginning of 2020, the world's population comprises about 7.8 billion people. By the year 2050, it is expected to have grown to just under 10 billion. This may sound a bit scary to those who still have a figure starting with 3 in mind from their school days. Fortunately, however, this does not spell the end of humankind, because objectively speaking, the "average human" living on this planet now has a better life than those who came before. Whether the non-human inhabitants of this planet are doing better now is, however, somewhat doubtful.

For the following calculations, let us assume a figure of 8 billion people. Now let us give each person a space of 10 m^2: 80 billion m^2 are 80,000 km^2, which is about the surface area of Austria. Not that bad, right?
Now let us form a human chain. Let us say that the participants hold each other by the hands and a person takes up about 1 m. This yields 8 billion metres or 8 million km. The human chain circles around the equator a whopping two hundred times! Or, if we were to use it to cover the distance to the moon (which is at a distance of about 400,000 km), we could travel ten times to the moon and back.

The world's population increases every year by about 75 million, which equals almost the population of Germany – just to give you a "net figure". We already know that a year has 30 million seconds. This means that with every second (!), three new earthlings see the light of day.

Let us now have a look at the flip side of the coin: Over the past decades, we have accumulated 8 billion tons of plastic. And the number is growing rapidly. It is well-known that a lot of plastic can be found floating on the surface of the sea. So, it must be lighter than water. 8 billion tons are equivalent to a volume of about 12 billion m^3. 1 km^2 is made up of 1 million m^2, which means that we could pave an area of 12,000 km^2 with a one-metre thick crust of pressed plastic, or the entire surface area of Germany with a layer of 3 cm. The "Great Pacific Garbage Patch" is currently floating between China and the USA. It has already reached a multiple of the surface area of Germany, but, following the above considerations, its average thickness must be less than 1 cm. What a relief ...

Caution is called for whenever large numbers are mentioned and it can't hurt to make some rough calculations. For instance, in October 2010, the dam of a tailing pond in Kolontár (near Ajkai in Hungary) burst. The pond was full of toxic aluminium sludge. News of the dam failure made headlines across the world. The following information was circulated by the media (and it can still be found on the Internet): An area of 40,000 km^2 was supposedly flooded with sludge (which would equal amost half the area of Hungary). The truth is that "only" 40 km^2 (still an impressive figure!) were covered in about 3 cm of sludge. What difference do three zeros make? Politicians mix up decimal places on a daily basis.

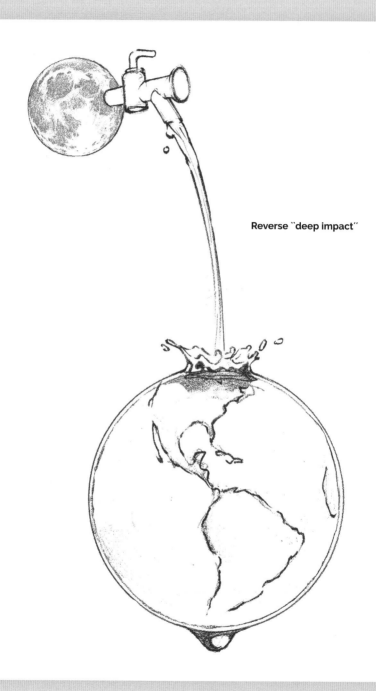

Reverse ``deep impact''

If the moon formed part of the earth ...

Let us imagine the moon to be entirely made up of water. If we pour this giant sphere of water (its diameter being about $1/4$ of the earth's) over the oceans, by how much would the sea level rise?

A spontaneous reply to this question might be: yet again, a useless example made up by mathematicians to torture other people. But you will see that it is actually quite interesting in more than one way.

Once (more than four billion years ago), the moon did indeed form part of the earth, until a giant impact (some theories suggests that it may have been up to 20 impacts) struck the earth. This certainly raises an interesting question: how much bigger was the earth before this "deep impact"? From a mathematical perspective, this question amounts to the same as the supposedly pointless problem that has been raised at the beginning of this page. (It has recently been proven that there is actually water on the moon, though only as frozen ice trapped under the surface).

Let us stop here to consider one thing: the oceans cover almost $3/4$ of the earth's surface and their average depth is about 4 km. We can now make the following rough assumption: if the earth had no land mass and were covered with 3 km of water, we would get approximately the same mass of water. So, returning to our example of the moon being poured over the earth: by how much would the earth's water level rise? The answer to this question might be difficult to estimate, but with some mental arithmetic, we should be able to come up with a solution:

Since the moon has $1/4$ of the diameter of the earth, the moon's surface is $\frac{1}{4^2} = 1/16$ and its volume $\frac{1}{4^3} = 1/64$ of the earth's. The volume of the earth would thus rise by the factor $65/64 \approx 1.06$, that is, by 6%. The radius of the earth's globe would grow by $\sqrt[3]{1.06} \approx 1.02$. (If you are good at mental arithmetic, you won't need a calculator: The third root will cause the surplus of 6% to shrink to about one third.) That is, the earth's radius would increase by 2% or $1/50$. We are taught at school that the earth has a radius of $6,000$ km. $1/50$ of the earth's radius equals 120 km. A rough estimate then suggests that the oceans would still (or "only"?) have a depth of about 123 km.

So, during the creation of the moon, 120 km were removed from the original size of the earth – mostly material from the planet's less dense crust, which explains why the moon has only $1/81$ (and not $1/64$) of the earth's mass.

While the surface gravity of an astronomical object on the one hand increases with the third power of the radius (mass increase), on the other hand, it decreases with the square of the radius (distance from the centre of mass). One would thus expect the moon ($1/4$ of the earth's radius) to have $1/4$ of the earth's surface gravity ($1/4\ g$). However, due to the moon's density, which is about $1/4\ g$ lower than the earth's, its gravitational force is only $1/5\ g$.

In hot pursuit

The right approach can save a lot of calculation

While making calculations, your common sense may come in handy. Some mental arithmetic can get you surprisingly close to your goal (or perhaps slower than you expected, as in the following, first example ...).

Example 1: Two joggers A and B run at a typical running speed of 3.5 m/s (which yields a time of just under 5 minutes per kilometre). Jogger A suddenly decides to return home, but jogger B is more ambitious and aims to reach a goal 300 metres ahead before running back.

B knows that, in order to catch up with A, he will have to run faster once they separate. So, he starts running at a fairly brisk pace of 4 m/s (it thus takes him only 250 seconds, that is, about four minutes, to run a kilometre). "See you soon", are his parting words. But what does "soon" mean in this case?

There are complicated ways of solving this mathematical example, but with the right approach you can turn this calculation into a "one-liner": B gains 1/2 m every second and it thus takes him 1,200 seconds to catch up with A. So, in this case, "soon" means 20 minutes!

Example 2: How long does it take a car driving at 108 km/h to overtake a slower car driving at 90 km/h, and how many metres must it cover to finish this manoeuvre?

The cars' driving speeds differ by 18 km/h, that is, 5 m/s (we must divide by 3.6, because we first multiply by 1,000 and then divide by 3,600). The average length of a car is 5 m. A car should pull out 15 m ahead of the front car and then return to the correct lane after another 15 m. This means that the driver needs to drive 35 m in order to overtake a slower car. It will take him seven seconds to do that. During this period, the faster car (driving at 30 m/s) covers 210 m.

Example 3: How much is the (negative) gravitational acceleration that you must overcome when you jump into a swimming pool from the 10-metre diving board?
For the first 10 metres, you are exposed to 1 g and if you jump head first straight into the water, you will plunge no deeper than 4 m. This means that you must decelerate 2.5 times faster (-2.5 g). Acapulco divers jump into only 3.6 of water from a height of 35 m. In this case, deceleration amounts to -10 g.

Having only 0.7 m of crumple zone after 20 metres of free fall (-28 g) can be compared to driving a car against a wall at 72 km/h. Fastening your seat belt won't help you much in such an accident. The wings of some insects must overcome up to 300 g. However, these wings carry no vital organs.

At least on earth, being in free fall has the effect of zero gravity, and air resistance will make your hair stand on end.

One year in free fall

L et us stay with large numbers for the following calculations: Occasionally, you will hear people say that "the economy has been in free fall for several months". How fast would we fall if, over the course of a year, we were to accelerate by 10 metres per second every second (which equals the gravitational acceleration on earth, that is, 1 g)? 30 million seconds with an increase in speed of 10 m/s per second yields 300,000 kilometres per second. So, we would be falling at the speed of light!

S uch calculations are, of course, purely theoretical, because, as we know from physics, we can never reach the speed of light, let alone exceed it. Neither is it possible for a space probe to actively accelerate for a year, because it would soon run out of fuel. It is, however, possible for probes to "work" with the gravitational force of celestial bodies (a process that is also known as *swing-by*). It is no coincidence that the Voyager 1 and Voyager 2 probes were launched in the year 1977. During this year, there was a particularly suitable constellations of the giant planets Jupiter, Saturn, and Mars, which the probes could use to gain additional acceleration.

B y now, the probes have crossed the boundary that marks the end of our solar system. At the moment, there are barely any forces acting on the probes aside from the latent gravitational pull of the black hole at the centre of our galaxy, but this force is compensated by the orbital rotation around the galactic centre. The outermost giant planet, Neptune, is 4.5 billion kilometres away from the sun. While sunlight reaches Neptune after only a few hours, it took Voyager 2 about 12 years to get there. After Neptune, things start to get pretty lonely ...

I f you jumped from a hot-air balloon flying at a height of 40 km, the atmosphere would barely slow you down. After 35 seconds in free fall (1 g), you would reach a speed of 350 metres per second and fall faster than the speed of sound.

If a rocket accelerated at 10 g for almost two minutes (which would really push the limits!), it would fly at the magical rate of 11,200 m/s (which is about 40,000 km/h), that is, the speed that is needed to leave the gravitational field of the earth. (We will not go into detail about the way this can be calculated, though it would be a comparatively uncomplicated undertaking.)

T he same can be observed the other way around: meteorites and asteroids are prone to hitting the earth at similar rates, because the process can also be reversed. In this case, potential energy is converted into kinetic energy: it is thanks to such an asteroid strike that human beings are populating the earth now. (We would not have stood a chance while dinosaurs were still living on earth, but they were wiped out after a massive impact 66 million years ago.)

Do these have anything in common?

What is the weight of a ladybird?

The volume of a cube is calculated by multiplying its length with its height. A cube with a side length of 1 cm thus has a volume of 1 cm³. A cube that is ten times as large already has $10 \times 10 \times 10 = 1,000$ cm³. The volume thus increases with the third power of the scaling factor. If both solids have the same density (e.g. that of water), the same holds true for their mass and therefore also for their weight. And when we think about the fact that *any* solid can be approximated by means of tiny cubes to a certain degree of accuracy, the statement thus holds true "for all solids that are similar to each other".

Let's google a few figures relating to large sea mammals: common dolphin: 2 m, 100 kg; blue whale: 30 m, 200 tons. With some imagination, a blue whale can be said to look like a (slim) enlarged dolphin. In its largest possible version, such a blue whale can be 15 times as long as a dolphin. That means that it would have to weigh $15 \times 15 \times 15 \approx 3,000$ times as much, and thus over 300 tons. The scale of this quick estimation sounds about right, considering that whales are indeed more slender than dolphins. Let us now put this statement to the test with a heavier specimen, the humpback whale (14 m, 30 tons): here, the scaling factor compared to the dolphin is 7. With the calculation $7 \times 7 \times 7 \approx 350$, we get 35 tons. Not bad.

The default values (length 2 m, mass 100 kg) could be compared to human data, even though a human who is two metres tall would have to be quite slim in order to "only" weigh 100 kg. What would then be the weight of a (slender) 1.50-m tall human? The scaling factor is 3/4; raised to the power of three, we get 27/64, and thus a bit less than half of 100 kg: 45–50 kg. This also sounds reasonable. The average baby measures about 50 cm and the scaling factor is 1/4; the infant's mass should thus be only 1/64 of 100 kg, and thus not even 2 kg. OK, the scale is about right, but here, we apparently have to consider that babies have much shorter legs compared to adults.

Let us go one step further and test the formula with a non-mammal that has a reasonably comparable shape: what is the weight of a honey bee? Its body length is about 15 mm – it is thus 100 times smaller than our 150-cm human. We would theoretically expect it to have a mass of about 50 mg. Effectively, bees have a mass of around 90 mg. They also have shorter legs, which means that a comparison with an infant (50 cm/4 kg) would yield more accurate results. We have to admit: the formula seems to work quite well ...What about an elephant (4 m, 6 t)? Let us stretch out its legs backwards (6 m) for easier comparison. It would already look less plump that way – almost like a sumo wrestler. Let us first compare it to our two-metre-tall basketball player from before: three times as long equals about 30 times as much weight, and thus 3 tons. We should keep in mind that the basketball player also weighs only about half of a two-metre-tall sumo wrestler.

Homework: the shape of a ladybird can be compared to that of an elephant. Measure its length and determine its weight ...

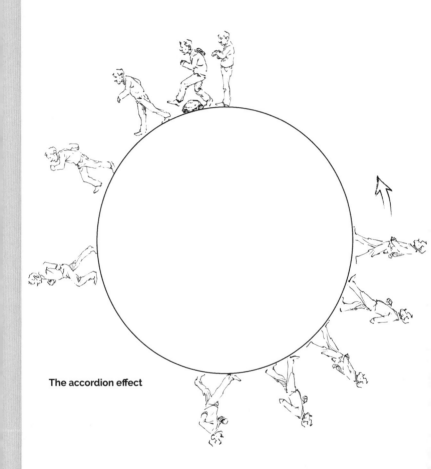

The accordion effect

Traffic jams and other choreographies

Consider a typical situation on a motorway: five cars driving at 130 km/h want to overtake two lorries travelling at 100. The driver of the second lorry hasn't looked in the rear-view mirror and pulls out to overtake the slowcoach before him. It takes the driver of the first car only half a second to react and lower his speed, but during this half-second he was driving too fast and now he must slow down to well under 100 km/h – to, say, 90 km/h – to avoid an accident. The second car driver sees the flashing brake lights of the car in front and hits the brakes after another half-second. Since he was driving at 130 both during the reaction time of the front driver and his own, he must drop his speed to about 80 km/h. You can deduce now what the third (slowing down to 70 km/h), fourth (60 km/h), and fifth (50 km/h) have to do for this scenario to end well ...

If all ends well after the second lorry has overtaken the first one, everyone will curse 'that stupid lorry driver' but it won't take long for the same game to start over again. Since there is *plenty* of room in front of the first car now, the driver accelerates to 130 again. The second driver, who had to slow down to 80, has to accelerate even more in order to close the gap in front. The third driver does not want to lag behind; so, he makes the most of his engine's horsepower to increase his speed back to 130. The engine of the fourth car roars just as vigorously. And the fifth driver puts his car to an even more extreme test by accelerating from 50 to 130 within a mere 10 seconds ...

What lesson can we take away from this mundane game? Should the five cars have driven at a lower speed? No, this was not the problem (they were all driving within speed limit). The problem was not their speed but the fact that there was *not enough distance* between the vehicles – but then again, the appropriate minimum distance will depend on the speed limits. Overtaking manoeuvres are not that dangerous as long as there is plenty of space to speed up, because then you won't have to brake hard if the vehicle in front suddenly pulls out to overtake.

This accordion effect can also be observed with military columns of marching soldiers: when a group of soldiers move through difficult terrain at too close a distance, the people in front will be able to walk at an even pace, whereas those at the back will continually have to slow down to a halt and then speed up again. The solution to this problem: leave more space in front of you!

Birds and fish also try to leave a consistent, preferably small gap when travelling in flocks. This gives them protection. Where there is much potential prey, predators are not far behind. When a flock is under attack, the individual that is closest to the predator will spot the danger first and flee hastily. The rest of the flock might not have realized that they are in trouble, but they have been programmed by nature to follow their fleeing mate. However, this instinctive reaction is always slightly time-delayed, which causes an accordion effect throughout the flock. The possible consequence: no leader, and hence no choreography.

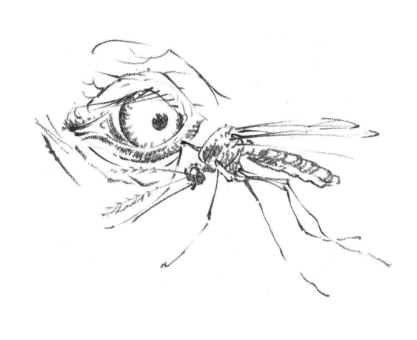

4

Contradictions

Δεν μπορώ να το κάνω!*

Εάν δεν το σκεφτόταν
τόσο πολύ,
δεν θα ήταυ πρόβλιμα.**

* I can't do it!

** If he didn't think so much,
   it wouldn't be a problem ...

Isn't it paradoxical?

A paradox is a statement that runs counter to an expectation or that seems contradictory. The "scaling paradox", which we will deal with at a later stage, belongs to this category.

Here, we are dealing with a question that allegedly caused sleepless nights for the classical Greek philosopher Zeno of Elea 2,500 years ago. The question is quite simple: a tortoise is moving at a constant speed of, let's say, 1 m/s (this number is, of course, a bit high, but this need not concern us at this point) and has a headstart of 10 metres. A speedy runner (in Zeno's example, it is the famous Achilles) moves ten times as fast. When will our hero catch up with the tortoise?

The problem can be solved quickly using "modern computational techniques". However, the ancient Greeks were fond of describing thought processes at length. There's obviously a snag in the following description: when the runner reaches the point where the tortoise had started from, the tortoise will have moved along another 1 m. When our sprinter reaches this new position of the reptile, it will have moved along another 0.1 m. When he reaches this point, it will have moved another 0.01 m, etc. We can repeat this ad infinitum – the four-legged friend will always be ahead of the runner. Does this mean that the tortoise can never be overtaken?!

The mistake in this line of reasoning, in fact, lies in the conclusion – even an infinite number of sentences can only describe a very limited time frame: $1\,s + 0.1\,s + 0.01\,s + \cdots = 1.1111\,s$. What happens *after* this time frame is another story ...

A typical paradox of daily life that has possibly lead to many a debate amongst neighbours with varying degrees of knowledge in the field of physics is the problem of removing moisture from a basement:
The always cool basement of a house has too much moisture and everything that one stores there begins rotting sooner or later. The neighbours who are not well-versed in physics think that the problem is best solved by airing the basement on particularly warm days – after all, warm air absorbs more moisture than cold one, and a blow-drier works with hot air.
This well-meaning initiative, however, makes matters worse: the warm air, which had sufficient time to absorb a great deal of humidity outside, now enters the cold basement, emitting moisture.
Some housemates who are resistant to advice, however, tend to ignore such explanations. Waiting for days on which the air outside is colder than the basement air is apparently too much to ask ...

If we walk up any higher, we won't even
have to cook these eggs!

Eggs on the Himalaya

It takes three minutes to cook a soft-boiled egg and ten minutes to get a hard-boiled egg that could be used as an Easter egg. Right?
I'm afraid it's not quite so simple. We must consider at least two components that play an important role here:

First of all, the size of an egg plays an important role. Have you ever bought a box of eggs (that were not previously sorted by weight) from a local farmer? If you have, you may have noticed that the size of eggs can vary considerably. The variation is even more pronounced when you consider their volume (which increases by the third power). Heat can only be transferred through the surface, which is enlarged by the square of the scale. This means that an egg with 1.5 the diameter of a smaller egg will take 1.5 times longer to boil to the same consistency than the smaller egg. An ostrich egg, which measures three times the diameter of a regular egg (hence, having a 27 fold volume and a 9-fold surface area), will take three times as long to reach a similar consistency in boiling.

Though not heated to boiling point, bird eggs must be kept warm ($37-38°$) in order for the embryo inside to develop properly. Ostriches leave their eggs to be matured by the sun's heat in the hot semi-desert, while smaller birds apply enough heat on their eggs by just sitting on them. Birds can have a body temperature of up to $42°$. Slightly above that temperature, the proteins in the brain could be damaged.

Second, the boiling point of water is dependent on the ambient air pressure. (Liquids begin to boil when their vapor pressure exceeds the surrounding pressure.) The air pressure at sea level is defined as 1 bar. In Mexico City or Addis Abeba, we have only 0.8 bar. The highest permanent settlement in the world (a community of gold miners in Bolivia 5, 500 metres above sea level) has only half the sea-level pressure (1/2 bar). In the death zone of Mount Everest (over $8,000$ m), the air pressure amounts to no more than 1/3 bar. Moving for several hours at such heights requires additional oxygen.
By the way, oxygen becomes dangerous under increased atmospheric pressure. Hans Hass, an underwater diving pioneer, almost had to pay for this with his life. Hass's diving activities raised awareness of the safety hazards of oxygen, and since then it has become common practice in scuba diving to calculate the partial pressure of the oxygen in the breathing gas.

Let us return to the low air pressure high up on the mountains: at 0.8 bar, water boils at slightly over $90°$. At 1/2 bar, boiling occurs at around $80°$. And at 1/3 bar, it takes only $70°$ for water to boil. So, at $70°$, you will have trouble cooking a soft-boiled egg, let alone a hard-boiled one. If you move even higher (for instance, 20, 000 metres high in a hot-air balloon), you could even eat an egg without cooking. However, you would have to make sure to wear a protective suit, because at such great heights your own blood could start to boil despite having a temperature of merely $37°$ ...

Ouch!

Making an elephant out of a fly?!

A re you familiar with the famous surreal painting by Salvador Dalí, "The Temptation of St. Anthony"? The temptations are sitting on giant elephants and horses that have extremely long, spindly thin, mosquito-like legs. This is reminiscent of the German saying that somebody who is being overly dramatic about a minor affair is "making an elephant out of a fly". The corresponding English phrase is "to make a mountain out of a molehill", but that image doesn't serve our purposes in quite the same way.

R eal elephants have *very thick* legs; with mosquitoes, it's quite the opposite – even though their legs appear quite thick when they are bloated with blood. These dangerous beasts (mosquitoes are some of the few creatures that deserve this label) are able to absorb 50 mm^3 of blood, in which case they have a mass of 50 mg at a body length of 1.2 cm, and thus a multiple of their "dry weight". A large elephant is 400 times as long as a mosquito (scaling factor $k = 400$) and should have a mass of $50 \cdot k^3$ mg = 3.2 tons while sporting a similar constitution (the volume and thus the mass – if the density is comparable – increases with the cube of k). The density of the body is comparable – indeed, the blood of the mosquito may originate from the involuntary donor.

H owever, why do the giant's legs need to be this thick, or, to put it differently, how does the mosquito easily manage to stay in balance with just a fraction of their cross-section? Through measurements, biologists have discovered that muscular strength does not increase proportionally

with the volume but is dependent on the cross-section, which, as we know, only increases with the square (k^2). The larger animal would thus only have $1/k$ of the strength in its legs if it had a mosquito's legs. This was, of course, all the same to Dalí – and rightly so: This absurdity is one of the factors that make his painting unique.

In order to have the same strength in its legs as the mosquito, the elephant's legs would need to be 20 times as thick in relation ($20^2 = 400 = k$). That sounds about right, having compared the elephant to the bloated mosquito, which after all is flying around with a multiple of its own weight.

S peaking of flying: would Dumbo really be able to fly (at least as clumsily as the bloated insect)? You're probably guessing no. After all, a mosquito flaps its wings a few hundred times per second. Dumbo's wings (but not his ears) need to have a sufficiently large surface: it also needs to be 20 times as large. At some point, a physical limit will be reached.

A justified question arises: why are Dumbo's "relatives", jumbo jets, able to fly (even without flapping their wings)? The secret lies in the velocity: "Jumbo" is only able to fly once it has accelerated to about 300 km/h. Starting with this speed, another paradox comes into play: the aerodynamic one (see also p. 63). If we somehow manage to make the airflow on the top side of the airplane wing bigger than that on its underside, the pressure drop lifts the 300-ton giant into the air almost effortlessly.

A smaller ball
must also have
less air resistance!?

A question of size: air bubbles and comets

The first thing you are told when you do diving course is: the most important thing is never to ascend too quickly. The apparently constant ascending velocity of small air bubbles is a practical maximal standard measure in order to gauge the speed of one's ascent. The next time you go diving, you will see that large bubbles will soon disappear from your visual field! That's because you have much more volume and push back much more water, which leads to buoyancy, according to Archimedes. It makes no difference that smaller bubbles have less water resistance, as that increases only with the square of the ball's radius (two-dimensional cross-section), while the (three-dimensional) volume increases with the third power (all of this still goes back to Archimedes). Because this entire process happens rather slowly (every three seconds, the bubbles have moved a metre further upwards), this is quite easy to check. The speed is a maximum speed because the buoyancy is the same as the resistance of the water.

On land, we have two problems: on the one hand, we learned at school that almost everything falls equally fast in a vacuum, whether it be a bird's feather or an iron ball. On the other hand, we have considerably less resistance, which means that balls and pellets accelerate so quickly that we soon lose sight of them. The maximum speed (a balance between weight and air resistance) is only reached much later and is much more difficult to capture subjectively.

The question whether a small ball drops faster than a big one made from the same material is nevertheless related to the aforementioned air bubbles: the larger one's fall is quicker because the cross-section of a sphere increases more slowly than the volume (and thus the weight) when it is enlarged. However, these differences only play a marginal role in the beginning (about the first 10 metres). Where this aspect comes into play is the moment when the air resistance counteracts the weight! In that case, the small ball will be considerably slower. For the same reason, water bubbles in a drizzle fall much slower than ordinary raindrops.

To take an extreme case, let us compare a small iron ball with a comet. While the iron ball (depending on its size) will not reach a speed of over 30 m/s, air resistance is irrelevant to the comet, and it crashes onto the earth's surface with a speed of several kilometres per second.

As a final example, let us compare a large ball made of a light material (e.g. wood) and a smaller ball made of iron. If the wooden ball is not large enough, the higher density of the metal ball (mass = volume × density) will provide a crucial "turbo" moment to the latter, which means that it will have a higher final velocity. However, in the end, the size does become relevant, and after a certain limit value, the wooden ball is unmatchable in terms of speed.

How does this heavy item stay in the air?

The Need for Speed

Everyone has, at some point, held their palm out from the window of a moving car and noticed the strong forces that are at play there when changing the angle of incident. This experiment is quite risky at high velocities, as your arm would immediately be yanked backwards as soon as your hand is upright. However, this still does not explain why an airplane weighing several tons stays airborne: if its position is too upright, it will come crashing down.

When thinking about the term *aerodynamic paradox*, you may remember an experiment from your youth that you will be able to reconstruct at any time. Hold a piece of paper in front of your mouth as horizontally as possible. The part that you are not holding will inevitably hang down flaccidly. However, if you blow on it *at the top*, the paper will move upwards into a horizontal position – as long as you don't run out of breath. If you want to perfect this experiment, you could use a blow dryer – the piece of paper will stay in the air much longer this way. These appliances have a much higher airflow velocity at the top than below (where it is, in fact, zero), and a higher airflow velocity creates a pull – just like a hurricane, which has no trouble uprooting trees and lifting houses as if they were weightless.

This is where explanations of the mechanics of airplanes usually end with the following sentence: due to the form of its aerofoils, the airflow that passes the wing has a longer path – and thus a higher velocity – at the top than below, which results in a pull owing to the paradox.

This may be true. However, the wings of today's airliners have an almost symmetrical cross-section and still manage to fly safely. It seems that the explanation based on a "longer path at the top" is not quite sufficient.

What is missing from the abbreviated explanation is the *sharp edge* at the end of the profile. When air flows around such an edge at a high velocity, air vortices arise behind it. Wherever there is a vortex, there is a counter vortex. Experiments showed that this counter vortex starts underneath the aerofoil and gets to the top by moving around the rounded off anterior end. This slows down the speed below and accelerates it at the top: there is, thus, *pressure* from below and a *pull* from above. That is what keeps the giant airborne. A certain minimum speed is always needed.

A helicopter (with a similar profile to that of an aircraft) gains speed from the rotation of its rotor blades; birds and insects do this by buzzing with their wings (the smaller they are, the higher the frequency). This also works under water: just think of the pectoral fins of oceanic sharks, which would sink if they did not actively swim. In water, which has a higher density than air, vortices are formed much more easily and even leisurely swimming creates sufficient vortices on the trailing edge of the animal's pectoral fins.

Catch me if you can!

A magic ribbon

What was the line in Goethe's *Epirrhema*?

Take in nature-meditation,
Each and all in contemplation,
Naught is inside, naught is out,
For the inside is without.

Let us take a rectangular, not too wide strip of paper and sketch the centre line with different colours or different types of lines at the front and the back. Now let us rotate one end of the strip by $180°$ and then join the two ends. The centre line is closed in this way, and the different colours (or types of lines) from the front and the back come together. We now have a classic *Moebius strip*, which has one impressive feature: it is "not orientable", i.e., it is impossible to distinguish between its inner and its outer surface. If we had not sketched the centre line on both sides in different ways, we would not have been able to see the transition between the two sides of the ribbon. The left edge of the upper side is at the same time the right edge of the underside, and the edges blend into each other seamlessly. Pretty confusing, right?

Foxes are well-known for their cunning in traditional lore, while rabbits are associated with being swift. Will the fox in the picture ever catch the rabbit? If both of them run at the same pace, the rabbit will remain unharmed anyway – the two are always diagonally across from each other. If the rabbit is quicker than the fox, it will at some point outstrip the fox, but neither of them will notice this, as they are running on different sides of the ribbon. At some point, the rabbit will increase its lead over the fox so much that it will end up on the fox's side – with the latter, however, *in front*. Because the rabbit is probably constantly looking backwards in panic, all the fox has to do is wait for it …

Would you like to stun your audience? The Moebius strip can also function as a "magician's tool": if you carefully scratch the original strip of paper in the middle in advance using a cutter, you can easily separate the two ends of the strip afterwards. You then get a strip twice as long and half as wide, which is "curled twice". The new strip now has two clearly differentiable sides – as opposed to the Moebius strip.

Because it's so easy, you could now go back to making a Moebius strip. This time, however, you furtively manipulated the strip you are using in such a way that you have split it into thirds, scratching it twice in parallel to the centre line. After gluing the ends together, you separate the strips again. You are left with a narrower Moebius strip, but there is now a twice-curled strip interlinked with it that is not a Moebius strip.

5

Integrals

Hey you – can't you integrate properly?

Differentiation or integration?

Here's another (abridged) mathematical joke: E-to-the-power-of-x seems to be quite bored at a party organised by Sine. No wonder – he cannot integrate. Cosine, on the other hand, hangs out with everyone (and everyone tries to get him to integrate a bit).

The joke, of course, only makes sense to those who have been taught integral calculus at school (who are becoming fewer and fewer): they will know that

$$\int \cos x \, dx = \sin x \ \text{bzw.} \int e^x \, dx = e^x.$$

Those who did not get the joke might learn a lesson from it – after all, this book is not only aimed at mathematicians but at everyone fascinated with maths.

First, let's cover the basics: differentiation and integration are inverse operations that can cancel each other out – a bit like addition and subtraction or multiplication and division. That's why

$$\int \cos x \, dx = \sin x$$

automatically leads to

$$\cos x = (\sin x)'$$

when we differentiate both sides of the equation.

The above joke demonstrates a number of things: the exponential function e^x is of special significance, because it is the only function that remains unchanged when it is integrated or differentiated:

$$\int e^x dx = e^x, \ (e^x)' = e^x.$$

For this reason, e^x could easily be termed "the queen of functions", even though she is a guy in our case.

The ominous number $e = 2.71828\cdots$, named after Leonhard Euler, plays as central of a role in mathematics as $\pi = 3.14159\cdots$. The latter is used, among other things, to calculate the circumference of a circle, or the volume of a sphere.

Euler's number e can also be found in a number of important formulas. Yet, it usually requires more advanced knowledge, because it is often found in conjunction with the so-called imaginary unit $i = \sqrt{-1}$. The following "beautiful" relation, for instance, applies:

$$e^{i\pi} = -1.$$

As a final note, E-to-the-power-of-x need not bother integrating: since

$$e^{ix} = \cos x + i \sin x$$

holds, he can accommodate both Cosine *and* Sine.

By the way, there is no reason for Cosine and Sine to pretend that they are different: aside from being phase-shifted, the two look exactly the same.

How am I supposed to finish this by the evening?

The problems of a genius

Gottfried Wilhelm Leibniz was a contemporary of Newton's; he had his productive period around 1700. And just how productive it was indeed! In fact, he only had two problems: first of all, his chronic lack of time. After he had thought out new amazing things during the night, he had to work hard during the next day in order to commit these thoughts to paper in an orderly fashion. This went on for 365 days each year. Allegedly there are still a number of notes by Leibniz that have not been reviewed in their entirety.

In case one of our readers could say the same about themselves – that is a common trait in productive people. Do you occasionally think something along the lines of "Newton is dead, Einstein is dead, and I'm also starting to feel a bit queasy ..."?
Are you missing the second problem? That reminds me of another joke: "There are only two kinds of mathematicians: those who can count to two."

Here's Leibniz's second problem, then: there was the infamous priority dispute with Newton. It has to be said that in Leibniz's and Newton's time, when the Internet or similarly quick communication devices were not yet available, it happened quite often that two exceptional talents worked on one and the same problem without the other person knowing. Certain things seemed (and still seem) to be in the air, resulting in different approaches to the same problem.

But what was the problem, really? Newton has already developed the main features of infinitesimal calculus in 1666 – however, he only published his findings more than 20 years later. Leibniz developed a concept in 1675 that we still use today. At the time, the matter was puffed up quite a bit in the world of mathematics and there were actually accusations of plagiarism. Nowadays, we know that the two of them simply arrived at comparable concepts independently. The fundamental theorem of calculus is also referred to as the theorem of Newton–Leibniz.

Leibniz was one of the last universalists and not just a "maths nerd". He was one of the most important philosophers of his time and was seen as a pioneer of Enlightenment thinking. Leibniz was also involved in the field of biology: as a pioneer of paleontology, he already discussed ideas of species change based on evolution a century before Darwin. Leibniz also communicated with more than $1,000$ people and wrote about $15,000$ letters, which form part of the UNESCO documentary heritage today. He probably did not have much of a work-life balance but geniuses like him luckily don't seem to be overly concerned about this aspect of their lives.

**Four apples, two and a half pears,
and the Riemann zeta function**

How old is the captain?

Everyone knows a joke of the following type: "A ship is sailing with 30 km/h from location A to a harbour B that is 120 km away. Having travelled half of the way, it has to reduce its speed to 20 km/h, in order to be able to reach its original speed afterwards. How old is the captain?"

The fun part about this is the completely unexpected and far-fetched question at the end, after you have already invested some time and effort into giving the correct reply as quickly as possible.

Maths students seem to find the following still more amusing (and people who don't know one or two technical terms may still join in the laughter): "John and Susan go to the supermarket and buy 6 apples and 4 pears. Upon leaving the shop, Susan gives an apple and a pear each to a homeless person. John eats one apple and half a pear on the way home. How big is the integral of the Riemann zeta function along the unit circle?"

Some entirely unexpected questions can, however, be solved by seemingly confusing preambles: "A bear roams around, moving 20 km to the south, then 20 km to the west, and finally 20 km to the north, eventually reaching its original point of departure. What is the bear's colour?"

After a moment's hesitation, it becomes clear that the bear must have started at the North Pole and we are, thus, dealing with a polar bear.

Let's try something else:
Think of a random number between 1 and 10. Multiply this number by 9 and calculate the sum of the digits of the product. Now subtract 5 from this sum. The ensuing number corresponds to a letter in the alphabet (the number 5, for instance, corresponds to E). Now think of a fruit and a country in Europe that does not have a border with Switzerland. Now for the question: What do dates have in common with Denmark?

A mathematician can only go back to his or her routine after such a question once he or she has proven that this calculation always yields the number 4 (and thus the letter D). This means that the respondents have just a few options, as hardly anybody thinks of a fruit other than dates and (in English) Denmark is the only country starting with D in Europe.

It thus suffices to demonstrate that at least the first ten multiples of 9 have the sum of digits 9, which – minus 5 – yields the desired number 4. Rather than making use of the theory of numbers in its entirety, why not simply put this claim to the test by trying it for the numbers 9, 18, ... , 81, 90?

Sine curves and
logarithmic spirals

The length of a curve

When a moth flies around a lamp in spiralling motions, or a drunkard moves waveringly from side to side on his/her way home, how long is their path?

Whether we are talking about spiral curves or sine curves, a curve in the mathematical sense is only partly related to the way that motorists use this term. Any curved line or curve is a *one-dimensional object* in mathematical terms. If a "zero-dimensional dot" rides the rollercoaster in such a curve, it, of course, travels a certain distance. That distance is called "arc length".

A circle is probably the first curve that we spontaneously think of. The formula for calculating the circumference of a circle is also well-known to anyone who has successfully completed six years of schooling – it immediately pops up without us giving it much thought: $U = 2r\pi$. Actually, we should say $U = 2\pi r$, as 2π is a constant and r is the variable radius (we also say $2 \cdot r$ rather than $r \cdot 2$). Upon hearing the number π, some people's eyes light up: everyone is utterly fascinated by the fact that there is this incredibly important number with an ominous value of $3.14159...$ where every subsequent decimal place is unpredictable. The number was first identified by Archimedes. However, there exists no exact construction for π, even though the ancient Greeks may have wished for it – as opposed to this, $\sqrt{2} = 1.414...$ is, for instance, very easily constructable as the diagonal of a square

with side length 1, even though the subsequent decimal place is equally unpredictable in this case.

Which other curve has an easily applicable formula for its arc length? The list is surprisingly short! One of the very few curves – but a fine specimen nonetheless – is the logarithmic spiral. We can spot it in a number of ornaments, or when we look at a snail shell from above. Theoretically, it winds again and again around its centre, and even if we keep using increasingly stronger magnifying glasses, the pattern will always stay the same. Despite these at least theoretically *infinite* windings, the overall arc length of this curve is *not infinite*. If our zero-dimensional dot is thus driving in its one-dimensional tunnel at a constant speed, it will still reach its goal quite soon, despite endless windings. In the case of the butterfly, the journey is over even earlier, because the lamp does not permit most windings in the vicinity of the centre in terms of physics. Butterflies indeed often fly along logarithmical spirals as a way of observing a constant course angle to the light rays in order to fly in an allegedly straight line.

The sine-shaped path of our drunkard is, by the way, only about $1/5$ longer than it would be on a straight line. This can, however, only be calculated approximately, by approximating the sine wave by means of a few hundred short paths. That procedure has to be done for almost all curves, in the absence of easy formulas ...

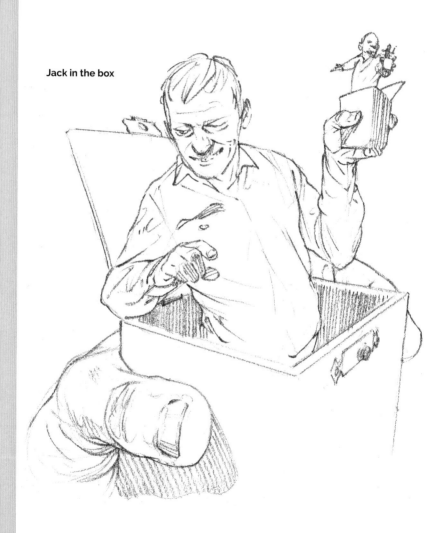

Jack in the box

Recursion and potentially infinite loops

The picture on the left will of course remind you of the cover of this book. That was our original drawing on the topic of "infinite loops", until we decided to incorporate Einstein on the cover. That way, our "Jack in the box" is additionally memorable – after all, Einstein himself was known for his humour!

Speaking of (seemingly) infinite loops, it is also noteworthy that Einstein had to wait a long time for his Nobel Prize. In the memorable year 1905, the then 26-year-old "technical expert third class" wrote as many as four groundbreaking papers while working at the patent office in Zurich; the most famous of them is, of course, the special theory of relativity. However, because members of the Nobel Prize committee contested the theory for some years, thus preventing Einstein from receiving the award, the scientist did not become a Nobel Laureate until 16 years later. The prize was awarded for another paper from the 1905 series, which meant that the infinite loop finally had an exit condition.

Imagine that you didn't know what an infinite loop is. You are most likely to check the index, hoping to find another entry that would explain more fully what is meant by this term. Wouldn't it be a shame if all you found was "infinite loop, see *recursion*" and "recursion, see *infinite loop*" …?

Recursion derives from Latin and means something like rewinding or retracing. In programming, recursive procedures belong to the most elegant but also to the most difficult to understand procedures.

Let's use the Fibonacci numbers as an example once again (see also p. 117). If we know two consecutive Fibonacci numbers – e.g. f_1 and f_2 – we can determine the subsequent one as the sum of the two numbers: $f_3 = f_1 + f_2$. Now we have a new pair of consecutive numbers, f_2 and f_3, and the next one can be found in the following way: $f_4 = f_2 + f_3$, etc. This is the only possible way of calculating these numbers, as there exists no formula that could be used for this procedure.

Another example, the *Koch snowflake*, is defined recursively in the following way: take an equilateral triangle. Now divide each side of the triangle in three parts of equal size and put another equilateral triangle on the middle third. We now do the same at every part of the outline, adding equilateral triangles at every side of the newly formed construct in the same way. The matter already becomes quite elaborate after only a few recursions, with the outline looking increasingly complex. It is even possible to show that its length becomes infinite. This leads us to the topic of the subsequent double page: fractals.

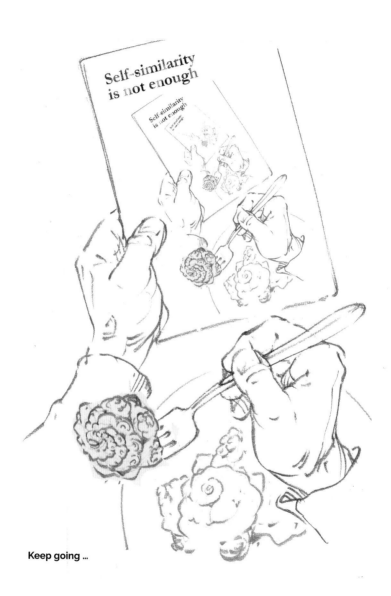

Keep going ...

What exactly is a fractal?

Until the year 1974, nobody knew what a fractal was. The term was coined by the mathematician Benoît Mandelbrot, who uses it to label certain naturally occurring or artificial objects and geometrical patterns. The term initially only seemed to find application in relation to programming gimmicks, but then Mandelbrot published his work "The Fractal Geometry of Nature".

A basic property of a fractal is "self-similarity", as well as "scale invariance" (that is, you can zoom in any number of times and the fractal should still look the same at each scale). Classic examples of a fractal in the strictly mathematical sense are the so-called Koch snowflake (see the previous double page) and Mandelbrot sets. Purely mathematical formulas also allow to create figures bearing a deceptively close resemblance to ferns.

It is quite remarkable how a fractal's shape keeps reappearing at whatever scale it is viewed. Per definition, it is impossible to determine the entire perimeter of a fractal unless you content yourself with the expression "infinitely long". In this regard, mathematicians make the following distinction: "ordinary curves" have a length that is measurable and they are classified as "one-dimensional". Fractal curves have a dimension somewhere between one and two (since their shape never fills the whole surface). Just to give an example, which you need not understand: the Koch snowflake has a dimension of 1.26, which is quite close to 1 but can still be difficult to grasp.

If we decide to be more tolerant and require only *a certain degree* of self-similarity and scale invariance, we will find a lot of fractals in nature: the shape of coastlines and rivers; the branching of plants, blood vessels, and lung alveoli; and even the distribution of star clusters across galaxies may qualify as natural fractals. If we magnify, for instance, the frond of a fern, we will see that each frond is made up of small frondlets that more or less correspond with the overall shape of the fern.

However, if we zoom in on these frondlets, the shape will already start changing, and as we keep magnifying, the self-similarity will disappear. In fact, it cannot be any other way, because otherwise the fern would have an infinite perimeter. The same holds true for clouds: Big clouds are often made up of smaller clouds that have a similar shape, but if we zoom in further, what we will get to see instead is droplets of water. Such shapes are still quite fascinating, and it seems fair to speak of nature as having a "fractal character".

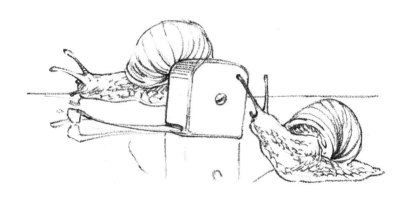

6
Measurements

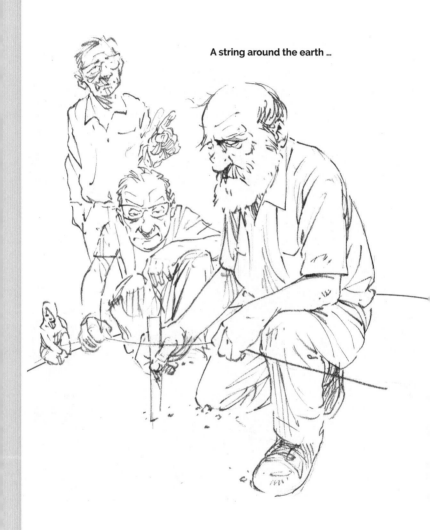

A string around the earth ...

40,000,001 m

Until not too long ago, the metre was defined as one ten-millionth of the distance between the equator and the North Pole. Accordingly, the earth's meridians had a length of exactly $40,000,000$ m. Due to the slightly imperfect spherical shape of the earth, the equator is 21 metres longer than a meridian. The earth is, in fact, (though still roughly speaking) an oblate ellipsoid. The bulges at the equator are a result of the earth's rotation around its own axis, and they are also one of the causes of the earth's wobbling motion ("axial precession"). Over the course of $26,000$ years, the earth's rotational axis, which appears to be fixed in space, completes a rotation whereby it traces out a cone whose axis is perpendicular to the earth's orbit.

Today, a metre is defined in terms of the constant speed of light in a vacuum, namely as exactly $1/299792458$ of the distance travelled by light in a second. Of course, it is also possible to say that the speed of light equals $300,000$ km per second. To the reader's surprise perhaps, a light year is then not interpreted as a unit of time but as a unit of length defined by the *distance* travelled by light in a year. You will have no trouble calculating that a year has 30 million seconds, which means that a light year equals 9 trillion km.

A light year is a useful unit when measuring space. While it takes only 8 minutes for light to travel from the sun to the earth, the closest star is already 4 light years away. The brightest star (Sirius) lies at a distance of 8.6 light years. Most stars that can be seen with the naked eye are no more than a few hundred light years away.

But let us return to the equator with its circumference of $40,000$ km. Let us assume that the earth is a smoothly polished sphere without any oceans and that there is a string tied around the equator. Now we will lift this string from the earth's surface in a uniform manner, thus making the string longer. Here's an estimation question: if the string is exactly one metre longer than before, how far away is it removed from the sphere's surface. If you do not know the answer, your estimations will probably be far too low.

This is how you calculate the answer:
We want to determine the new radius. We get the radius by dividing the circumference by 2π. For $U = 40,000,000$ m, we get an earth radius of $R = U/(2\pi) = 6,370$ km. If we extend U by 1 m, we get an additional $1/(2\pi) \approx 0.16$ m, that is, 16 cm. This is probably more than you estimated, right? And you will realize the following thing: no matter how long the initial radius of the earth, even with a tiny sphere the string will be lifted 16 cm away from its surface if we extend it by one metre ..

In Syene and $5,000$ **stadia to the north ...**

Measuring the world

The globe of the earth is double-curved. Every intersection of the globe with a vertical plane will yield a circle that has the same radius and circumference as the equator or the meridians joining the two poles.

There were already some clues hinting at the spherical shape of the earth in Antiquity: navigators who sailed along the West-African coast towards the equator made reports of an everchanging sky, with the stars continuously changing their positions. A contemporary of Archimedes, Eratosthenes of Alexandria, calculated the earth's radius in the meridian more than 2,200 years ago in the following manner:

The ancient city of Syene was situated almost exactly under the Tropic of Cancer (near the modern-day city of Asswan). Back then it was perhaps widely known that if a person looked down a well in Syene at noon on the summer solstice, the shadow of the person's head would block the reflection of the sun in the water, because the sun's rays fell vertically into the well. Through astonishingly precise measurements, Eratosthenes was able to determine that at exactly the same time in Alexandria – 5,000 stadia (800 km) to the north of Syene – the sun's angle of elevation (measured with a rod perpendicular to the earth's surface) equalled $1/50$ of a full $360°$ angle (today we would say $7.2°$). This led him to draw the correct conclusion that the meridian must be 50 times as long as the 800 km distance between Alexandria and Syene, namely 40,000 km ...

When we measure longer distances (for instance, the trajectory of an airplane from A to B), we look at the vertical plane that runs through A and B. On this plane, which, in addition to the points A and B, also includes the centre of the earth, we will find another "great circle" (with a circumference of 40,000 km). As can be proven with mathematics, the length of the shorter arc is the shortest distance between A and B. So, airplanes should generally fly along this route, unless there is jet stream (with suitable tailwinds) nearby, which might be more convenient in terms of fuel consumption.

The shortest distance between two points thus cannot be found along any of the earth's circles of latitude (the only exception being the equator): if a billionaire is suddenly overcome by an urge to travel from Kuwait (on the $30°$ latitude line) to Houston, Texas (also on the $30°$ latitude line), the airplane will not fly straight west, but far to the north. This will result in a shortcut of 1,450 km, which means that the billionaire could reach his destination by direct flight in a jumbo jet. However, whether landing is permitted or not might depend on the respective president in office ...

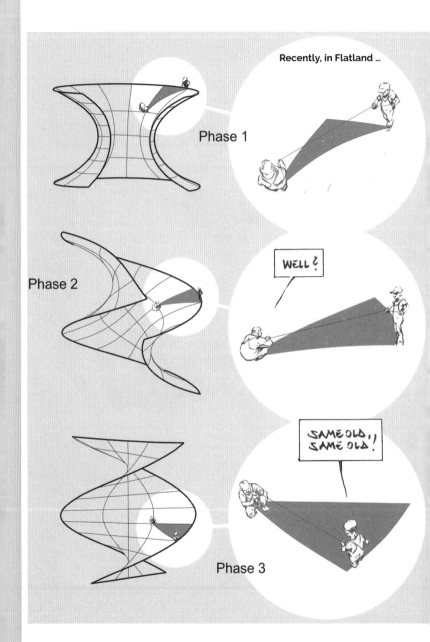

Exciting surfaces for minimalists

There is something special about spheres. Regardless of the way you slice them through their centre, you will always get the same shape: a great circle. Any other planar intersection of a sphere will also yield a circle. Mathematicians have even gone as far as to prove that you will also get a circle when the sphere is not touched by the intersecting plane – the intersection curve is then simply an "imaginary" one. For a sphere to be created in space, a celestial body must have a diameter of more than 500 km. These bodies usually have a liquid core, which is then formed into a sphere by gravity. There are also spheres with smaller dimensions (for instance, dew drops and soap bubbles), but these spheres emerge for very different reasons: surface tension forms bodies with minimal surface area and maximum volume.

If we dip a wire hoop into a bath of soapy water and pull it out again, we will create bubbles, which, for the fraction of a second, will assume a fairly stable shape, a so-called minimal surface. Bubbles are the only *enclosed* surfaces of this kind; no other surface encloses a volume, but their shape is still determined by the force of surface tension.

The theory of minimal surfaces with no volume can be quite a challenge to mathematicians. At least the following things can be said with certainty: let us assume P to be any given point on a surface, while E is a plane touching the surface containing a plane P. Then E touches the surfaces after two curves that intersect each other vertically in P. Now this is just the beginning: once we have found a minimal surface, we can bend it in any direction while still retaining its minimal qualities. We can thus create a whole set of surfaces out of one surface.

What follows is a typical mathematical line of thought: since we cannot see with our naked eyes that we live on a sphere, it is perhaps not far-fetched to imagine intelligent creatures living on a minimal surface. These creatures measure their world just as we do: they measure angles and "trajectories" (distances) that are generally curved (without them noticing). You can see this in the drawing on the left side: the creatures do not notice that the surface that they are carrying changes as they bend it because there is no change in the angles and distances! What may sound like a fantastic scenario at first can actually be observed in nature: when seaweed or soft corals sway with the current of the ocean, their relatively stable surfaces create a close approximation of a minimal surface and they will effortlessly bend without straining or buckling. …

One way ticket

Ants have optimisation problems, too!

Imagine being a busy politician, campaigning and attending events all around the country. No day goes by without having to travel from A to B to C etc. How can your journey be optimized? If you travel to, say, four different destinations, this optimization might still be done "manually": you know all the distances involved in your journey – namely, AB, AC, AD, BC, BD, and finally CD. As you can see, optimizing a journey with four destinations can already seem quite a demanding task: if you always start your journey at A, there are six different ways in which you could travel to the other destinations: ABCD, ABDC, ACBD, ACDB, ADBC, ADCB. However, calculating six sums and then selecting the shortest journey should not pose much of a problem.

Now imagine having to travel to 20 or 30 different places. The complexity of your optimization task would downright "explode". You would always have to compare the sum of all distances and then select the minimal value. Now you might obviously ask, "Why not use a computer?", but it turns out that even the best computers would be completely overwhelmed with this task if it involved even a few dozens of destinations. So, computers cannot offer us an easy way out once we have to cover a higher number of destinations. That is why many mathematicians, physicists, computer scientists, and people from other professions have already racked their brains over "the travelling salesperson problem", trying to find a better solution. They have come up with a few tricks that make the problem at least more manageable.

One of the more original versions has been inspired by the behaviour of ants. Individually, ants might only have very limited intelligence, but their strength lies in connecting thousands of information pieces into a "hive mind". But how does this really work?

Let's say there is a stream of ants swarming from a starting point P towards a destination Q in order to reach a new source of food. Insects like ants and bees possess a remarkable gift: they can, "in principle" maintain their direction of travel even if they have to walk around several obstacles. If ants encounter an obstacle, they will simply move around it in a random way. During their travels, they will leave a slowly fading trail of scent. So, between points P and Q, there will be many pheromones that will only dissipate after some time. The trail of those ants that have randomly managed to find the optimal path will be the one that has faded the least. So, if other ants of the colony wish to travel from P to Q, they will only have to follow the strongest scent trail. Occasionally, they might "mis-smell" what way to go, but their journey back will be much easier. And this is how it continues: the more attempts the ants make to travel from P to Q, the more efficient will be the ant trail established to join the two points in an optimal way.

Teaching a computer how to "sniff out" the optimal path would, of course, be quite a challenge ...

Captain, our pendulum clock has stopped working!

What's the time?

G PS is a relatively new tool. Before that, navigation required considerable skills and resources - be it on a ship, while travelling on land, or even on an airplane. Think of the two expeditions to the South Pole in the austral summer of 1911/1912, when a group of Norwegians led by Amundsen and a group of British explorers led by Scott tried to reach the southernmost tip of the earth in order to stuck their national flags into the eternal ice by travelling completely different trajectories. To Scott's dismay, Amundsen's flag was already there when the British group finally arrived. However, this shows that both groups managed to navigate their way to the South Pole with considerable precision. In order to find his way back, Amundsen had built snow cairns, where he deposited information on the mound's exact position, as well as the direction and distance to the previous cairn. Scott had better done the same - his team froze to death during their return journey, only 18 km away from their base camp and supply depot.

H ow is it possible to determine one's position on the earth without GPS? If your destination is the South Pole, you can just head to the south following your compass while it still works. As you get closer to the pole, you will no longer be able to rely on a compass because the magnetic south pole is actually some distance away from the geographical pole. The angle of elevation to the celestial pole (in this case the south celestial pole) - which can be found using the stars while there is no continuous daylight for 24 hours - will give you the latitude. When the sun is shining while you are exactly at the South Pole, its disc will appear to stay at the same angle of elevation for 24 hours. Amundsen and Scott had to measure the angle of elevation several times a day, using a sextant with an artificial horizon. The results had to be continually readjusted and verified when the weather allowed for it.

H owever, determining the latitude is easier than figuring out the altitude of your current location, especially during explorations across the infinite waters of the oceans. While there were only pendulum clocks, which would usually break during severe storms, it was not possible to make statements such as the following: when I left Santiago de Chile ($71°$ west) heading towards the west, the sun reached its highest point at noon. Now it reaches its highest point at 10:00 a.m. This means that I must have travelled 2 time zones, which equals $30°$. This means that I must currently be at $101°$ west. I still have to travel $8°$ to the west to reach the Easter Island. Should I be mistaken, I will probably starve to death or die of thirst, because there is no land within a radius of $1,000$ km.

O ver many decades, the finest brains tried to come up with an elegant yet practical solution to this problem until, finally, a clock was invented that was not likely to break with the rocking movements of a ship …

What?! We're supposed to be going straight on? But we're in front of an abyss!

Let the computer show you the way!

Nowadays, we often turn to GPS ("Global Positioning System") for help. That is highly practical, after all. Even if I get lost in the woods - one of my apps will surely get me home safely. Nowadays, even taxi drivers in big cities punch in street names into their phones before we're off. Interestingly enough, airplanes do not rely only on GPS; they do additional checks using conventional methods, despite the fact that GPS works particularly well in lofty heights – there are no mountains, trees, or buildings.

In the 1990s, the American military shot 27 satellites into well-defined orbits at a height of 20,000 km 'the flight altitude thus amounts to three times the earth's radius, and only geostationary satellites are still further up than these). For a few years, the signals that they emitted were artificially corrupted, so that only those institutions that could decode these deliberate mistakes would be able to navigate entirely accurately. Afterwards, a few mathematicians came along, developing methods by which the encryption could be automatically removed, and eventually, it was abandoned altogether. If the latest reports are to be believed, still other mathematicians have found methods for manipulating the signals in such a way that users think that everything is alright – while being on the wrong path. Missiles that are allegedly capable of distinguishing between a men's and a woman's toilet could thereby be led to detonate somewhere safe.

Is it possible to explain how GPS works in just a few sentences? Well, first we need three satellites A, B, and C. If we have the appropriate signals supplying their current time down to a nanosecond and their current positions down to a metre, we can work out our own position if we know our own time as follows: the signal is transported at the speed of light. The difference in nanoseconds until the signal arrives is given accurately to a metre by the distances a, b, and c from A, B, and C. If we now intersect three spheres A, B, and C with the radii a, b, and c, we get two points of intersection. One of these points will be quite far away in space or below ground level, the other one will be our own position. If we have a fourth satellite, the matter is unambiguous.

The difficulties lie in the detail, of course: firstly, we can "only" rely on time designations obtained with the help of satellites down to microseconds, despite atomic clocks on board. Secondly, we generally do not possess atomic clocks ourselves (those are far from affordable!). However, if we repeatedly apply the described calculations, first assuming an "approximate" time (accurate to a ten thousandth), we will also be successful, as long as there are no troublemakers who invent new dirty tricks or the signals are not distorted in other ways – or, indeed, all electronic devices are incapacitated for some reason.

Planet gears

Pulleys and planet gears

According to the law of the lever, it is theoretically possible to lift all kinds of loads:
"effort × effort arm = load × load arm". A similar principle can be applied to other useful gadgets. Leonardo da Vinci drew a sketch of a pulley system, where you have to pull a fairly long rope on one side in order to slowly lift a heavy load on the other side. Here the following principle holds: "Force times distance remains constant". This is achieved through a combination of rope pulleys.

Planet gears work in a similar manner. They are used to convert fast rotations into much slower, yet more efficient rotations. A typical example of this are electric screwdrivers, which are powered by an electric engine with a relatively weak battery. This engine can move a tiny front gear at about 60 rotations per second. How can this be reduced to only one rotation per second in spite of the rotational force being 60 times more powerful?

Well, you could try putting some auxiliary gears between the tiny front gear and add a ring gear that is 60 times as large. Then the tiny wheel would have to make 60 rotations in order for the large ring gear to make one rotation. Or you could come up with an even better solution: how about dividing the process into two steps? As a first step, you could join a ring gear that is only 8 times as large with a new, tiny front gear lying underneath. This second tiny gear would then move another ring gear that is also 8 times as large, and this ring gear is connected to the screwdriver blade. This solution would yield a conversion ratio of $1 : 64$.

The last line of thought is quite typical of mathematicians, who like to boil ten litres of water by bringing only one litre to boil and then saying, "The rest works the same way" (this is probably one of those mathematician jokes to which other people are likely to respond with a weak smile).

You could easily join the second ring gear with another tiny front gear, which could, in turn, cause a third ring gear to rotate, and so on. The conversation ratio would increase exponentially and would then be $1 : 8^3 = 1 : 512$, and if you take it a step further, $1 : 8^4 = 1 : 4096$, etc.

At level 4, the blade would move at about only one rotation per minute, thus being rendered quite "unstoppable"!

7

Physical

Who is who?

$$1\,Pa = 1\,\frac{N}{m^2} = 1\,\frac{kg}{m \cdot s^2}$$

Definitions and units

Here's a joke told by physicists (abridged): Einstein, Newton, and Pascal are playing hide-and-seek. Einstein is the player designated to look for the others (also known as "it"); he isn't able to find well-hidden Pascal, but manages to find Newton, who has not been hiding at all, sitting on a carpet of one square metre. That way, Einstein has found Pascal after all.

In case you haven't laughed already: the definition of pressure as a unit in physics is 1 Newton per square metre and is called 1 Pascal. You are only familiar with the unit 1 bar or 1 at (technical atmosphere) or psi (pounds per square inch)? That's absolutely fine. You could always work with the conversion table for the most common units on Wikipedia.

Units aren't straightforward. Back in the day, the elder of the two authors of this book wanted to get a driver's licence in the US in addition to the European one he already had, because driving licences seem to be more important documents there than passports. The undertaking almost failed because he ticked the wrong boxes for the following multiple-choice questions:

• How many ounces of "86-proof liquor" (what the hell is that?) correspond to a six-pack of beer? (It turned out that 86-proof liquor is 43% schnapps).
• How long is the brake path when driving 55 mph (miles per hour): 143 feet, 243 feet, or 343 feet?

• A school bus stops in the opposite lane on a two-lane freeway. How fast are you allowed to drive? (0, 15, 25, etc.)

Said author is unsure of the correct answers to this day and has so far not met an American who knew them with certainty, but at least the first two questions are supposedly not asked today anymore.

Quite interesting in this respect is perhaps the correlation between mpg (miles per gallon) and litre per 100 km. For this, only the following conversion formula is helpful: in order to get one value, we have to divide 235 by the other one respectively. Thus, 6 litres pro 100 km correspond to an itinerary of 40 miles with one gallon (ca. 3.8 litres). In this case, we, of course, mean the international mile (1.61 km) and not a nautical mile (1.85 km).

The conversion from Celsius to Fahrenheit is not trivial either: from Fahrenheit to Celsius, you have to subtract 32 and then multiply the result by 5/9. The other way round, you have to multiply the Celsius value 9/5 and add 32.

Final question: how many zeroes does a billion have in German-speaking countries? – 12. That may explain the one or the other misunderstanding …

Any other suggestions?

Just ask the caretaker!

Here's, once again, a classic joke told by physicists – an anecdote surrounding the brilliant physics student (and later Nobel laureate) Niels Bohr. He once had to answer the following question in an exam: "Describe how one can determine the height of a multi-storey building with the help of a barometer." Bohr felt provoked – the solution using the difference in air pressure is, of course, obvious. However, it is almost too trivial for him and besides, it is nevertheless probably too imprecise. Flippantly, Bohr replies: "Let us tie a long piece of string to the upper end of the barometer and lower it to the ground from the roof of the building. The length of the string plus the length of the barometer will yield the height of the building."

The honourable board of examiners becomes a bit disgruntled, asking the cheeky devil to come up with an answer that would demonstrate his knowledge of physics better than the first one. Bohr proceeded to give a series of unexpected answers:

• One could throw the barometer to the ground and calculate the height by means of the drop time;
• One could – provided the sun shines – work with the length of the shadow thrown by the edge of the building;
• One could let the barometer swing as a pendulum, first below and then from above, on a long string: from the difference in the oscillating period, one could also derive the difference in pendulum length;
• One could also go to the caretaker and gift him the barometer in return for the correct answer.

You could say that silly questions usually lead to silly answers … If we are being honest, the questions that mathematicians ask occasionally seem quite far-fetched, and a hackneyed answer seems like "just punishment" for such a question. Generally speaking, a solution based on common sense and practicability is preferable. In fact, it should be a welcome challenge for scientists in their ivory tower to be able to "explain complex issues to their own grandmother". If that is not possible, integral calculus or something similar may need to be used. However, it is indeed the case that it is simply easier to explain some issues by avoiding overly scientific vocabulary.

Let's put the "grandmother theory" to the test: why is a sidereal day four minutes shorter than the average solar day? Let us use the fact that a full angle has $360°$ and a year has about 360 days. In order to be able to see the sun at its peak at noon, we therefore have to "overtwist" each day by $1°$. 24 hours = 24×60 minutes divided by 360 gives 4 minutes. That is why the stars appear in the sky 4 minutes earlier each night than the day before. You could also make a 20-page academic paper out of this, with a result that would be comparably accurate down to the decimal place …

Get the goldsmith!

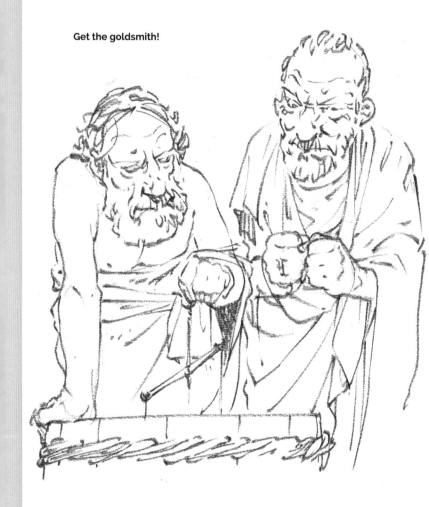

Archimedes and two fundamental laws

The Greek $A\rho\chi\iota\mu\eta\delta\eta\varsigma$ from Syracuse (Sicily!) reached the proud age of about 75 years before being killed by a Roman soldier, whom he had rebuked with the Latin words "Noli turbare circulos meos" ("Do not disturb my circles"). With his inventions, the genius had prevented the Romans from invading his hometown for three years.

Two of his inventions are well known among students today: the law of the lever and the principle of flotation. Both laws are often mentioned in connection with sayings that are famously attributed to Archimedes: "Give me a place to stand and I will move the earth with it" and "Eureka". The latter, which can be translated as "I have found it", was reportedly exclaimed by the genius as he jumped out of his bathtub and ran out onto the street naked – back then, this action may have caused less alarm than it would today.

By combining these two laws, the universal scientist uncovered a fraud in the manufacture of King Hiero's crown, which the king had ordered to be made of pure gold. Using the following method, Archimedes discovered that the goldsmith had cheated the king by adding less valuable (and thus also lighter) metals to the molten gold: he balanced the crown against a lump of pure gold on a scale and then submerged the scale with the crown and gold in water. Within the water, the side of the scale with the crown drifted upward.

According to the principle of flotation, the buoyant force that acts on a body submerged in a fluid equals the weight of the displaced water. (So, due to its small volume, a lump of gold experiences less buoyancy force than lighter metals.) The downward force on the body is its weight when it is not submerged and it acts on the body's centre of mass.

Archimedes and the tides

The gravitational pull of the moon attracts the water (and anything else) on the side of the earth nearest it. Since the earth rotates around its own axis, the point where the moon's pull on earth is strongest moves across the world over the course of a day. The following problem arises now: there are not just one but two high tides a day.

In order to come up with an explanation for the second tide, we must apply one of the principles of Archimedes: remember his saying "Give me a place to stand and I will move the earth with it." Let us now calculate the shared centre of mass that lies between the moon and the earth:

We will place 81 coins (the earth) on the left side of a scale and only one coin (the moon) on the right side. The pivot point of the lever must then lie at $1/82$ of the lever's length, which corresponds to a distance of about $400,000$ km. The shared centre of mass between the earth and the moon thus lies at a distance of $5,000$ km from the earth's centre in the direction of the moon, which means that it lies inside the earth. In addition to all their other trajectories in the solar system, the two celestial bodies orbit around this point in the course of about 4 weeks.

For once, we ask you to turn to the next double page to continue reading.

At least the ship has the right profile ...

Archimedes deserves a second double page

Here is the announced sequel to the previous double page. Describing the work of Archimedes inevitably raises a lot of talking points.

O bjects lying on the surface of the earth that faces the moon lie seven times closer to this shared centre of mass and they, therefore, experience only one seventh of the centrifugal force arising from the moon's orbit compared to objects lying on the opposite side of the earth. Now we must also consider the relatively fast rotation of the earth around its own axis, which affects the opposite side of the earth by producing a second non-trivial tidal wave that also moves around the planet.

Archimedes and the profiles of ships

Think of the profile of a ship's vessel. It is designed in such a way that when the ship tilts, the volume of the displaced water, which generates buoyancy, will have a centre of mass that lies "on the right side" in relation to the ship's own centre of mass. The buoyant force acting upward and the gravity force acting downward on the ship's (hopefully stable) centre of mass generate a torque that moves the ship in an upright position. Suitable ship vessels must be designed in such a way that they can generate a ship-righting torque at *all angles of inclination*. For this purpose, the ship's centre of mass is best located as low as possible.

I f the centre of mass lies too high in the ship's vessel or if it relocates uncontrollably due to barrels or similar objects rolling across the ship, the situation can become quite dangerous. Professional sailors know how to use this to their advantage: they will lean to the left and then to the right, thus causing the centre of mass to relocate in order to tilt their boats to extreme angles.

Archimedes and volumes – sphere, cylinder, and cone

The versatile Greek was particularly proud of the following discovery: he managed to establish a relation between the volume of a sphere and the volume of a circumscribed cylinder (its meridian section being a square).

P eople are well familiar with this relation today because the formulas on which it is based are quite well known as being fundamental in mathematics: from the formula for the volume's cross-sectional surface, $Q = \pi r^2$, we derive the formula for its volume: $2r\,Q$. The volume of the sphere equals exactly $2/3$ of the cylinder's volume, and thus twice the volume of a double cone due to the top and bottom circles of the cylinder (with its tip lying in the centre of the sphere). The multitalent was also able to approximate the numbers $\pi = 3.1415\cdots$ to several decimal places. He did so by drawing a 96-sided regular polygon inside and outside a circle. These neat "boxes" then yielded the famous formula for a circle's circumference, $2\pi r$.

Fetch!

A return ticket to the stratosphere, please!

Everyone knows what a "trajectory parabola" is: when we throw a stone, its trajectory will be a parabola. For the first half of the trajectory, it will fly upwards, reach its peak exactly in the middle and then fall to the ground along a curve symmetrical to its ascent. That's theoretically how it works. However, let's not forget air resistance, which increases with the square of the stone's velocity. It slows down the stone from the first second onwards and its speed decreases as a result, which means that the stone eventually slumps to the ground at some point. That is called a third-order parabola. With this in mind, historical drawings that show the trajectories of cannon balls starting straightforwardly and then falling to the ground almost vertically in a circular arc are not as naïve as we may think at first.

Let us assume there were no air resistance (as is the case on the moon, for example). Is the curve a perfect parabola in that case? Well, a parabola is, broadly speaking, an ellipse whose focal point has slipped into infinity. Even without air resistance, this focal point is still the barycentre of the moon, according to Kepler's first law of planetary motion. The radius of the moon is $1/4$ of the radius of the earth, thus measuring $1,600$ km. Its flight parabola would then be (not just since Kepler) a flight *ellipsis* with a focal point that is $1,600$ km away from it. With a bit of imagination, this could be called a parabola – but only if we are speaking of "regular" throws, reaching up to a few kilometres. Alfie, our dog on the left, should be running the other way if the object is thrown extremely far.

If our goal is for the stone never to land on the Moon again, we would have to throw it vertically with "escape velocity". This velocity is about $8,500$ km/h. In order to fully get away from Earth, we need almost five times that speed. The solar system can only be left behind at a speed of a whopping $150,000$ km/h. Fortunately, our Earth speeds around the sun with an average speed of "only" $108,000$ km/h, moving exactly so quickly that it stays within its orbit and is not dragged towards the sun.

Now, let's turn back to air resistance. It decreases with height: with about every $5,500$ m of height, it decreases by a half. In practice, a commercial aircraft needs about 3 tons of fuel per hour at a cruising altitude of 3 km. At 12 km, it unfortunately still needs 2.5, because it flies faster – and time is money. That's why even airplanes on flights between Munich and Stuttgart fly as high as the stratosphere. Whether this not particularly large difference has any impact on the environment is another matter.

Another round!

Highway to the Danger Zone

Most satellites – around 2,000 – fly relatively close to the surface of the earth. That's quite a jumble, considering that their necessary orbital velocity is almost 30,000 km/h. Occasionally (though not very often), collisions occur, which has considerable consequences: after such a collision, hundreds of splinters fly around in space as uncontrolled projectiles with the same insane speed. With each further hit, the situation becomes increasingly messy. In 1978, the astrophysicist Donald J. Keller had already described this syndrome, which was then named after him: at some point, the zone in which the satellites and their splinters are speeding around becomes a danger zone that can only be traversed safely with a bit of luck.

Once you're out of the danger zone, there is nothing to see for a while. In a flight altitude of 20,000 km, about 30 GPS satellites are darting around – these are controlled and fly along precisely precomputed orbits. Their altitude is forty to sixty times as high as that of the others. It is highly unlikely that collisions will occur at this altitude.

Almost another 20,000 km further, we witness a bizarre situation. There are almost no satellites at that height, save for the space directly above the equator, where a sort of procession is taking place. At this altitude, there are about 300 "geostationary satellites", flying at a velocity of 11,000 km/h in a uniform orbit. They need the same time for a full circle as the earth needs for a full rotation (about 24 hours). That means that they always remain at the same point above the equator, fulfilling their devised function there.

Originally created as weather satellites, most of these are communication and television satellites today: who would be willing to live without thousands upon thousands of television programmes worldwide? At this impressive speed, the satellites usually have a distance of less than a thousand kilometres between them. If you could hover at a certain point in space as an astronaut (which you can't), such a construct would flash past every half minute. Finding a tiny bit of space for new satellites is thus becoming increasingly difficult.

'er the hilltops is quiet now, reads the opening of Franz Schubert's "Wanderer's Nightsong". Indeed, satellites are quite a rarity at a great height. However, there is a satellite at a flight altitude of about 400,000 km, and thus ten times as far away from earth as the geostationary ones. It has been there for a very, very long time – almost since the earth came into being. It also has ¼ of the diameter of the earth; our friend in space has a considerable effect on life on earth. Its face is always the same because it turns around its own axis as quickly as it turns around the earth. And for good reason: its self-rotation decreased as a result of tidal forces (which exist without water, too!) until this balance was eventually achieved.

8

Biological

In the beginning was an egg – laid by a non-chicken

What came first: chicken or egg?

The chicken-and-egg problem is applicable to many an everyday problem. The fascination with this philosophical question dates back to the ancient Romans and Greeks. A similar question frequently arises in escalating relationship spats and is usually answered in different ways. The question whether black holes existed before galaxies or vice versa is an example of a "cosmic chicken-and-egg problem". Physicists tend to say that black holes came first. In further consequence, there was a strong interplay between the galaxies and their invisible "turbo drives".

Laying eggs is a classic form of procreation: the egg contains all necessary matter for the offspring and very often, the parents do not look after them further. Temporally speaking, the egg, of course, came "before" the hen if we consider the fact that insects and fish were already laying eggs hundreds of millions of years ago. In between – and probably also before then – there were egg-laying reptiles and dinosaurs and there still exist a few select mammals that lay eggs). A valid question would thus be: did the Meganeura (a giant dragonfly) mother or their eggs come first? Biologically speaking, the question is not directly relevant, as neither the Meganeura nor their eggs fell from the sky but both formed part of a long evolutionary process.

As the procreation of a chicken also involves a rooster, one could also say (to make matters even more confusing) that the rooster came first! English geneticist John Brookfield formulated the matter like this: "In the beginning was an egg – laid by a non-chicken." In addition, it is important to note that it needs a second generation until the genotype for laying eggs catches on, because the genotype of an animal does not change during its lifetime.

Mathematicians take the easy way out. They refer to a sequence of occurrences for which predefined dependencies are fulfilled as *topological sorting*, saying succinctly and cryptically that "topological sorting cannot be applied to the hen-and-egg problem."

The geometry of a bird's egg is thus all the more fascinating, and indeed, academic papers have been written about it. The egg is rotationally symmetrical but it is not symmetrical in terms of an equatorial plane (as, for instance, an ellipsoid). This has the advantage that it cannot simply roll away, which would mean the death of the fledgling. Rather, the egg gyrates or wobbles, ideally coming to a standstill not too far away from the hen. Birds that hatch on steep cliffs occasionally display almost angular eggs, providing additional security.

That's got to impress her!

Pufferfish and symmetry

Pufferfish has a long tradition on Japanese menus. Nowadays, only specially trained cooks are allowed to prepare this fish: its venom belongs to the most poisonous ones in the animal kingdom – if you eat it. Dolphins, for example, were observed playing with puffers by pushing them around like a plaything, while at the same time being quite careful with them. It seems that they only want to sniff at the poison a bit, which provides them with a sort of high.

It was not until the 1990s that divers first discovered a strange behaviour in male puffers of the Torquigener genus: they build sandcastles measuring several metres in diameter, in order to impress females. These constructs are perfectly symmetrical and reminiscent of 24-hour clocks. If the whole thing had not been caught on video, some people would no doubt have recognised this as the work of extra-terrestrials wanting to send us a message; a fish, they would say, would never be capable of constructing such an aesthetically pleasing masterpiece, especially of this size. In human terms, we are speaking of structures with diameters of thirty metres, constructed without the help of a protractor or measuring tape. Completing such a sandcastle takes the fish an entire week of continuous work! The reward consists of the female fish laying its eggs precisely in the centre of the mating gift.

Symmetry plays a highly important role in nature – in the plant kingdom as in the animal kingdom. Let us mainly concentrate on the animals here:

• sponges, being the simplest multicellular animals, are not symmetrical at all.

• jellyfish, anemones, and other cnidarians are radially symmetrical – as is our pufferfish sandcastle. Radial symmetry can often be observed in the plant world, in the layout of blossoms and petals.

• fish – and more generally speaking, vertebrates – but also most other animal classes possess a symmetry plane (they are "bilaterally symmetrical" and thus have a "front" – the direction of movement). Even though an octopus has eight arms, it is nevertheless only bilaterally symmetrical.

• - Echinoderms (starfish, sea urchins), on the other hand, often display symmetry in five directions (animals with six, seven and many more arms developed from them in the course of evolution) – this is all the more impressive as their larvae are bilaterally symmetrical and go through a metamorphosis at a later stage. Even a number of interior organ systems are arranged in terms of five arms or symmetry axes. This development is unique in the evolutionary history of animals. Sea urchins can be thought of as throwing up their five arms upwards, and indeed, we can see muscular suckers in-between the quills – such suckers are also found on the underparts of the starfish.

An efficient model despite the simplification

Mathematical simulations of nature

Leonardo da Pisa, better known as Fibonacci, was the most significant European mathematician in medieval times. His knowledge was partly based on the work of Arabic mathematicians. "The son of Bonacci" had the following idea:

It takes a young pair of rabbits one month to reach sexual maturity. From that point onward, they can give birth to a pair of rabbits each month and this new pair will, in turn, reproduce after another month. If we were to conduct a monthly rabbit census, we would get the following results: 2, 3, 5, 8, 13, 21, 34, 55, 89, etc. A mathematician will, of course, immediately strive to establish a formula for this; however, that is not a simple task. First, one may prove that the number of rabbits established in a given count will consist of the sum of the two preceding results. Thereby it is possible to show that, as you go higher in the sequence, each subsequent number can be approximated very well by multiplying the preceding count by the so-called *golden ratio* $\Phi = 1.61803\cdots$. This number has a similarly magical ring to it as the number π, and nobody is able to predict what the next decimal place is.

When entering the results of the census in a Cartesian coordinate system, we notice quite soon that the points lie on a curve that quickly takes on very high values. After 20 months, we already have a figure of over 10,000 rodents, and after 40 months, hundreds of millions. This number is not as unrealistic as it might seem when one considers, for example, the feral rabbit problem in Australia. One thing is clear: At some point, the entire system is bound to collapse. In the case of the zealously reproductive, furry animals, this is often brought about by a lack of food and/or epidemic diseases, which may annihilate entire populations.

The following question, which you may be familiar with in one way or another, is comparable to the topic at hand – it also deals with *exponential growth*: A water lily population spreads across a lake in such a way that the area covered by water lilies doubles each week. After 20 weeks, the lake is entirely covered. At what point in time was it only half-covered? (You know the answer.)

If we think that such things can only occur in the realm of not particularly intelligent, instinct-driven creatures, we are in for a big surprise: The same reproduction curve demonstrably exists for humans. Unfortunately, our own curve has already reached a phase where its figures are vertiginous in relation to all other creatures in this world. Yet, there is hope that in the near future this rapidly rising population curve may level off sharply (we must be optimistic).

Comparisons on a small and large scale

People often think that the same phenomena may manifest themselves in small things as in great ones. This might be true in certain areas (for instance, the behavioural patterns of humans), but it does certainly not apply to nature – as we have already established several times. Especially the relation between a thing's volume (mass) and its surface area, or cross-sectional area, is dependent on its absolute size! This has, of course, manifold effects – both direct and indirect ones. Physicists have always wondered how, or under what circumstances, things could be compared on a small and large scale, including how this could be done in different "fluids", which is the term used in fluid mechanics to describe both gases and liquids (such as air and water).

How does the frequency of an animal's wingbeat or fin flap relate to their speed?
For this purpose, the physicist Vincent Strouhal took the quotient of the frequency divided by the forward speed and multiplied that value with the amplitude of the wing or fin movement (absolute size!). He thus came to the following conclusion: the highest level of effectiveness is reached around a value of 0.3. This number can be applied to the movement of both flying and swimming animals, irrespective of their size – that is, to animals ranging from mosquitoes over birds and bats to baleen whales.

Under which circumstances can, for instance, the flight of a tiny fruit fly be analysed on a larger scale? The physicist Osborne Reynolds calculated the quotient of the fictitious force and the viscous force, which act on any body in a flow of fluid. The resulting Reynolds number increases proportionally to both the size of the object and the flow velocity, and it decreases (in an inversely proportional manner) with the kinematic viscosity of the fluid. If we fabricated a hundredfold magnified model of a fruit fly, which beats its wings 300 times a second in real life, we could examine it in an oil tank, where it would only have to beat its wings once a second to "fly" within the same "Reynolds range". If the settings are adjusted correctly, the physical properties of the flow will be exactly the same in both cases so that we can draw conclusions from the model that would also apply to reality.

Here's a funny afterthought that still has a strong connection to physics: which animal takes less time to shake water from its wet fur – a dog or a squirrel? As you can probably imagine, the small rodent needs only a fraction of the dog's time. The time that each animal takes to complete this task depends on the angular velocities and the associated centrifugal forces. So, the rapid movements of smaller animals make it possible ...

A game of counting off

The golden evolution

Let's write down the Fibonacci numbers once again, where the next number is always the sum of its previous ones, which means that we don't have to learn them by heart: $2, 3, 5, 8, 13, 21, 34, 55, 89, 144, \cdots$. If we divide a Fibonacci number by its respective predecessor, we get a numerical sequence $3/2, 8/5, 13/8$ etc. that tends towards the famous number $\varphi = 1.618\cdots$, which has gone down in history as the *golden ratio*. Large Fibonacci numbers are thus the best integral approximations of the powers of φ. However, they need to be divided by the constant $\sqrt{5}$. Let's try it out: $\varphi^{11}/\sqrt{5} \approx 89$, $\varphi^{12}/\sqrt{5} \approx 144$ etc. Here, we can see clearly that we are dealing with exponential growth: the numbers become very large very quickly.

The alleged spirals found in sunflowers and daisies only partly provide information regarding the question how the flower heads really increase in size. Computer simulations show that the "golden angle" plays a decisive role in this process. Let us examine it in greater detail. Let us first comment on the number of spirals in flowers: we can always find "both" left-hand and right-hand spirals, and their quantity is always a Fibonacci number.

We arrive at the golden angle by dividing the complete angle of $360°$ at a ratio of $1 : \varphi$. This gives us $137.52\cdots°$ as a result. The computer simulation clearly shows that the aforementioned flowers arrange each following single flower in such a way that they follow from the preceding one, rotating by the golden angle and "slipping out" from the centre of rotation. Is the golden angle innate to nature? It seems the answer is yes – although it is debatable whether a higher power is responsible for this. It is more likely to be the result of a simply evolutionary process. If we try out the growth process with another angle, the amount of single flowers that can be accommodated is much smaller. However, single flowers mean progeny. And the progeny once again has the genes of its forebears. It is thus inevitable that those flowers that have approximated the golden angle most closely by means of random mutation will be the ones with the most progeny.

Finally, a hint for the newly enamoured: it may be of interest that two odd numbers always follow an even one. When counting the petals of daisies or sunflowers, and even roses, it becomes apparent that their quantity is a Fibonacci number (depending on their size). This is linked to exponential growth and the configuration of their single flowers. The probability for an odd number is thus 2/3, which is why it would in fact be better to start the game with "he (she) loves me not, he (she) loves me" rather than staying faithful to the original sequence "he (she) loves me, he (she) loves me not" ...

Eureka??

Fascinating diagrams

We already mentioned Archimedes and his legendary exclamation "Eureka". Perhaps the Russian mathematician Георгий Феодосьевич Вороной (Georgy Feodosevich Voronoy) had his mathematical epiphany while sitting in a bathtub and playing with soap bubbles – all of us have probably done such a thing in the past and we were fascinated with their geometry. Maybe it was then that Voronoy had the idea that one could display such shapes geometrically, in spatial diagrams of any dimension. If he had anticipated that computers would one day be capable of simulating such diagrams in the blink of an eye, he would probably have done the same as Archimedes and painted the town red (in his case Warsaw, where he was living at the time.)

What, then, is the definition of a Voronoy diagram? The best way is to think of a number of points (centres) randomly drawn on a piece of paper. If we take a closer look at one such centre C and its immediate neighbouring centres, we can see that there are regions on the plane of paper where each point has a distance to C that is shorter than the distance to all other centres. In this region, the influence of C is thus "dominant". Geometrically speaking, the sides of a polygon are formed by the middle bisector of two neighbouring centres respectively. These regions have quite a simple shape: they are so-called convex polygons, meaning polygons without indentations.

It is no coincidence that such polygons are reminiscent of, for example, dried mud, enlarged plant leaves, or a Somali giraffe's fur: if we imagine thin sludge (without cracks) drying up slowly, the evaporation will be stronger at some centres than at others, and tension will necessarily arise from these centres. The sludge will then start cracking "approximately in the middle" (and thus along the middle bisectors). With regard to fur, one can imagine that there are so-called inhibitors and activators that prevent and promote a certain fur colour respectively. In three-dimensional space, many formations – such as wasps' nests or basalt formations – resemble such diagrams quite closely, which suggests that there may be an underlying struggle going on with regard to spheres of influence that leads to impressive shapes and artistic formations.

Voronoy diagrams are impressive constructs in themselves. If we now allow the cell points to move towards the barycentre of the respective cells, we get new, refined diagrams. This process becomes stable quite quickly and the ensuing results are genuine masterpieces – optimised in many respects. It is highly likely that this constitutes an important principle in nature.

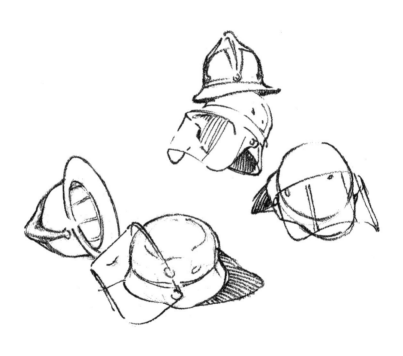

9

Statistics

**If you can do
THIS many press-ups,
you have nothing to fear!**

Muscles, statistics, and probability

Internists at Harvard Medical School in Boston have been analyzing the health data of 1,104 firefighters over a few years. In this study, they discovered that the number of press-ups that a middle-aged, active man is able to do give indications with regard to his heart health. This discovery led to many a clickbait-type headline (such as the one on the left).

The scientists discovered that subjects who were able to perform 40 press-ups were at a considerably lower risk of a heart attack, stroke, or other cardiovascular diseases. In the course of this 10-year study, 37 test persons suffered a heart attack – 36 of them by those participants who had managed to do fewer than 40 press-ups.

Well, it certainly sounds logical that sporty people should be less prone to heart problems. Let's put this idea to the test. As there are few people who are able to perform that many press-ups, let us assume the following for testing purposes: every 40th test person succeeds at reaching this limit. That means that around 28 subjects belong to this select circle, while the remaining 1,076 do not. One of those 28 athletic individuals develops a heart problem (the probability is about $1/28$, thus around 3.6 per cent). Of the remaining 1,076, 36 become problem cases – that's $36/1076$ and thus 3.3 per cent. Hang on. This figure does not correspond to the findings of the study at all!

New calculation, new chance: 54 test persons are top athletes (more than 5 per cent). That leaves 1,050 who are not in as great a shape. The probabilities for a heart attack now lie at around $1/54 \approx 2\%$ and $36/1050 \approx 3.4\%$ respectively. Now the results seem to fit. However, what if a second top athlete had been summoned to the cardiologist (which is a possibility)? That would mean $2/54 \approx 3.7\%$ vs. $35/1050 \approx 3.3\%$. A mismatch yet again ...

What we can learn from this is that one should be careful with premature conclusions. "Rare events" are full of pitfalls. A typical example of this is the discovery of rare infections and diseases.

Let's assume we had a very cheap and very reliable test (hit ratio: 99%) for a rare infectious disease. On average, every 1,000th is infected. We now want to carry out a test over a wide area. With 1,000 test persons, we will make a find once on average, but the test will nevertheless indicate that 10 persons are affected. These 10 unfortunates will have to come in again for post-tests. They should, in fairness, definitely be told that the probability for a false alarm lies at 90% – otherwise, there may be yet another top athlete suffering a heart attack ...

There is a
growing
number of yogis
in Germany

...and counting

Yoga is certainly healthy. One "representative BDY study for yoga in Germany (yoga in numbers 2018: interest is still growing)" shows the following:

In 2014, there were 2.6 million active yogis, and 9.6 million of the German population used to practice yoga. In 2018, there were 3.4 million active yogis and 7.9 million used to practice yoga.

It goes on to say that "the share of currently practicing yogis is considerably higher among women (9%) than men (1%)".

The last sentence is not unambiguous. Do they mean 9% of all 42 million women plus 1% of all 41 million men in Germany? That would yield a result of a bit more than 4 million, which is quite close to the suggested 3.4 million. This would also mean that 90% of currently active yoga practitioners are women, since the number of men and women among the population is almost equal.

In 2018, there were thus 0.8 million more *active* but 1.7 million fewer *former* yogis. Even if the entire increase of active practitioners had come from former yogis, 0.9 million would still be missing from the picture. We don't have to be worried about them, do we?

Some other statistics are no laughing matter – for instance, the one featured in the March 1th, 1979 edition of *New York Magazine*. The article deals with murders and subsequent death penalties committed by people of colour and by white people respectively. In the first case, 2.4% of trials ended with the death penalty, in the second, 3.2%. The average citizen may deduce from this that convicts who were white were judged even more strictly than those who were persons of colour.

However, further research shows that there was not a single death sentence for white offenders if the victim was a person of colour. Conversely, if the victim was white, every sixth offender was sent to death row. The seemingly higher rate of death penalty convictions among white murderers is thus the result of white people murdering white people, too. This is how an abridged truth becomes untruth.

Nowadays, we could formulate the matter as follows: short tweets containing alleged facts all too often fall under the category of "fake news". There has been a perceptible increase in the proliferation of misinformation. One famous example is the current president of the United States, who puts out a whopping 300 misleading claims a month on average according to the Washington Post (January 2019) to almost 60 million followers.

Let us conclude this spread with something fun: the "liar paradox". Everyone is familiar with the story of Pinocchio, who is occasionally a liar – he is easily exposed, as his nose starts growing every time he lies. Now think about this – what would happen if Pinocchio paradoxically said, "my nose is growing"?

We don't want Arabic numerals!!

The majority is against it

The phrasing of a question is decisive. Isn't it mean to ask whether Arabic numerals should be part of the curriculum while not adding that we are using these numerals for calculations on a daily basis? And what's more, the survey was published, too! To quote the Washington Post (17 May 2019): "A survey conducted by Civic Science, an American market research company, asked $3,624$ respondents: *Should schools in America teach Arabic numerals as part of their curriculum?* The poll did not explain what the term "Arabic numerals" meant. Some $2,020$ people (or 56 per cent) answered "no". 29 per cent of respondents said the numerals should be taught in US schools, and 15 per cent had no opinion." Civic Science gives a margin of error of three per cent.

To be fair, the numerals are written a bit differently in the Arab world, even though their meaning is the same. It is particularly noteworthy that zero is indicated by a dot, while the character that looks like zero to us denotes five. An American would no doubt recognize the number 6 as a 7 (see picture on the left). Once we are familiar with the conversion table, however, we can read Arabic equations just like European ones. These numerals were already used in the Arab world while Europeans were still struggling with Roman numerals (see p. 33).

Speaking of representative surveys, the number of respondents, and margins of error: let's say we want to predict the election result of a major party as reliably as possible down to 1%. For this, we need – believe it or not – $10,000$ respondents. We have not yet allowed for the fact that the selection of people and their honesty should be viewed critically.

Polling institutes merely survey between 400 to 500 persons before elections as a rule, which means that we can only expect the results to be accurate down to a few percentage points, give or take. However, these few percentage points – sometimes even just a few thousand votes – may be decisive in the race for the election winner. In that sense, such surveys are not particularly meaningful, and furthermore – interestingly enough – highly dependent on the polling institutes' preferred party.

Lastly, a remark about "resolute minorities": let's say that an association of 25 members lets their members vote on a number of issues via postal vote. Five members have secretly decided to vote against a certain seemingly inconsequential proposition, where other members have more or less randomly selected *yes* or *no*. We can now calculate that this minority will get their way with a probability of 87%, without making a big fuss about it.

Let me help you a bit ...

Don't you want to switch your choice?

You are close to winning the big prize (a sports car) in a game show. There is only one problem to be solved: You must choose one of three doors in a wall right in front of you. Behind one door is the car. Behind the other two doors are booby prizes (for instance, goats as in Monty Hall's show "Let's Make a Deal").

So, you pick a door, but instead of just revealing what lies behind that door, the game-show host (who knows where the prize is) will open one of the other doors to reveal a goat. You will be relieved, because your chances seem to have increased to $50 : 50$ (the other goat or the Porsche). But at this point, the host will ask you if you would like to switch your choice and open the other closed door. Is he trying to help you or is he only discouraging you from winning the big prize?

Since there is so much at stake, it is worth having a closer look at this problem. On the Internet, the "Monty Hall problem" is hotly debated. Consider the following line of thought: I first choose one of the doors at random. I will have picked either the door with the sports car or one of the two consolation prizes. The chance of picking the door containing the main prize is $1 : 3$. If I stick with my initial choice, my chance of winning will remain the same. The host's revealing of a goat does not have any influence on it either, because this will always happen no matter if you picked the door with the sports car or a door with a goat.

Let us now turn to the strategy of switching your guess. There are three possible outcomes: If your initial choice was the car, you will lose. However, if you have picked goat 1 or goat 2, your next pick will be the coveted prize *in both cases*. Seen from this angle, the information provided by the host is indeed helpful to me after all! Switching to the remaining door is unambiguously the better option.

If you discuss the problem in a small group, you will hear all kinds of contradictory arguments. Why not try the following: Prepare a game of three shells and a pea and play the game with a test person, telling them to stick with their *initial* choice twenty times and then *switch* their choice twenty times. The group will be astonished at how striking the difference between sticking with your original choice and switching your choice is, revealing switching to be the better strategy by far.

Maybe this time!

Chance has no memory

Millions of people worldwide gamble at roulette tables or place bets at the lottery week after week, thinking with the same bright optimism: "*This time*, I'll win. There has been a long streak of red, and the ball must land on 17 at some point". Now and then people do, of course, land a lucky strike, which is later fashioned into a beefed-up and exaggerated story.

Heads or tails? When tossing a coin, the following line of thought admittedly sounds quite reasonable at first: "The probability of getting heads three times in a row is $(1/2)^3 = 1/8$, and for four heads in a row, it is only $(1/2)^4 = 1/16$, which equals about 6%. So, if the coin has already landed on heads three times, there is a high probability (94%) that it will flip to tails next time." Perhaps this statement would not grab our attention if there was not so much at stake, but the stakes are actually quite high: namely, whether or not it is possible to "gamble yourself rich" systematically – many, many people have already racked their brains over this.

The fallacy in the above statement is quite clear: We do not bet on the entire series of coin tosses. We only observe, and when the coin happens to land on heads three times in a row, we are suddenly alert: "Quick, let's bet on tails!" But the coin has no memory. What's past is past. New toss, new chance. $50 : 50$. That's it!

Let's try something else: it has been proven that, after numerous tosses, the count of heads and tails will be almost the same. You might call it "restorative justice". Now let's apply this notion of justice to ourselves: "If I have already made three wrong bets, chances are that I will win next time." Well, as you will have realised: we are faced with the same dilemma once again. Why should fate care that you have just lost three times in a row? The above observation only applies after many, many coin tosses – say, $1,000$ at least. By the time you get to the thousandth toss, you might have already gone bust, and there is no point in getting angry with justice ...

Sometimes, there is a way of improving chances: let's say you play a game of "Russian roulette" – with a toy revolver, obviously, for safety reasons. There is only one bullet in the revolver cylinder, which has six chambers. If you give the cylinder a good spin every time before pulling the trigger, there is always a probability of $1/6$ that your opponent will catch a toy bullet. However, if you only spin the cylinder at the beginning, the probability will increase with every shot.

One more thing: should you really get ten heads in a row in a game of heads and tails, there is a good chance that the person tossing the coin knows some trick to manipulate how the coin is flipped ...

We're almost there!

Just you wait!

Think of a combination lock for a bicycle with four numbers between 0 and 9. Unfortunately, you have forgotten the code. Will you manage to pick the lock? After all, there are 10 possibilities for the first digit, another 10 for the second, etc. Thus, there are $10 \cdot 10 \cdot 10 \cdot 10 = 10^4 = 10,000$ possibilities in total ...

You will, then, have to approach this problem systematically (unless you have a certain talent for safe-breaking and notice when you have guessed a correct digit by being alert to certain audible subtleties – there are dozens of videos out there showing someone who only needs a minute to pick such a lock).

If the lock used letters instead of digits, of which there are 26 in the English alphabet, this task would be even more difficult. We would be faced with 26^4 (and thus almost $500,000$) possibilities. Still, even this code could be cracked, as long as we have a system. Theoretically – but really just purely theoretically – it is even possible to pick a lock with 100 or even more positions (letters). However, you would need a *very, very* long time – almost an infinite amount.

The question is: would we be able to crack the code even if we worked in a random way, without any system at all? The answer is: yes! At some point, sooner or later, we would get the right combination. That is "guaranteed" by the *law of large numbers*. It turns out that we will need the same amount of time, regardless of whether we are working in a systematic way or choose the numbers randomly. In a computer

simulation, the hard-working machine took a very long time to come up with the word "MONKEY" 100 times, but it needed about 300 million attempts to – entirely randomly – write down the 6 letters in the correct order.

This leads us to the *infinite monkey theorem*. The theorem states that a monkey randomly typing away on a keyboard may at some point type any book in the world. Alternatively, we could employ an infinite number of monkeys (with just as many keyboards) in order to reach the same result. Each new predetermined letter makes the problem even more complex, but that's the way it is in mathematics: the number of attempts is never truly infinite.

This entire matter sounds completely wacky and could be discounted as unnecessary shenanigans. The theory is irrelevant anyway, at least as far as the monkey is concerned. It is only deployed to explain (for instance, in the context of the theory of evolution) that a great deal can happen randomly when you have a great deal of attempts.

P.S.: Do you recognise the actor in the drawing? Maybe you've seen the movie by Stanley Kubrick, where he loses his mind in a hotel in the Colorado mountains, endlessly typing "All work and no play makes Jack a dull boy"?

It can't be true!

Aliens everywhere

"Those who know nothing must believe everything." This aphorism was coined by Marie von Ebner-Eschenbach but also served as the slogan of *Science Busters*, the Austrian equivalent of the American TV show *MythBusters*. It could equally be the slogan of the present book.

Occasionally, one meets people who would self-identify as "realists". They explain to you how David Copperfield causes objects of gigantic proportions to disappear or how Uri Geller manages to bend spoons. Occasionally, they have themselves mastered one or two magic tricks that are difficult to unveil, and still, there are things that they are unable to make head or tail of and where they seriously suspect that aliens must be behind it.

To be clear: the probability that extraterrestrial life already exists or will exist in the vastness of outer space is almost 100%, mathematically speaking. There are serious scientific theories stating that single-celled creatures may indeed have been brought to earth via a comet and that new lifeforms developed from them. When we look at the incredible results that evolution has produced on earth, it is almost unimaginable that the same does not happen a thousand times over in the universe.

However – once again, seen from a mathematical point of view: the probability of a temporal and spatial overlap between two civilisations that are so similar to each other that they could communicate tends towards zero. Our planet has existed for 4.5 billion years, while "civilised" mankind has only been around for 4,500 years (to choose a simple number) – a millionth of the timespan that the planet has existed. Even if there were a planet not too far away that had living conditions similarly friendly to life (such as Mars), a temporal overlap of highly developed civilizations is highly unlikely. If such a planet is dozens of light years away, the distance creates an additional problem – if our theories about the speed of light are correct.

In this light, it appears almost ridiculous how frequently an "inexplicable light" or something similar serves as proof that we are surrounded by aliens. Crop circles are a classic and well-known example that shows how easily people may be lead astray. In recent times, scientists have proven just how easy it is to "conjure" complex circle patterns on a corn field using very simple tools. Teams of dedicated volunteers made imprints of a range of predefined patterns into corn fields in front of a camera one night. What's the use? Someone always raises the argument that it must have been aliens after all. That's the way it is with humans – they have to believe in something.

10

Vision

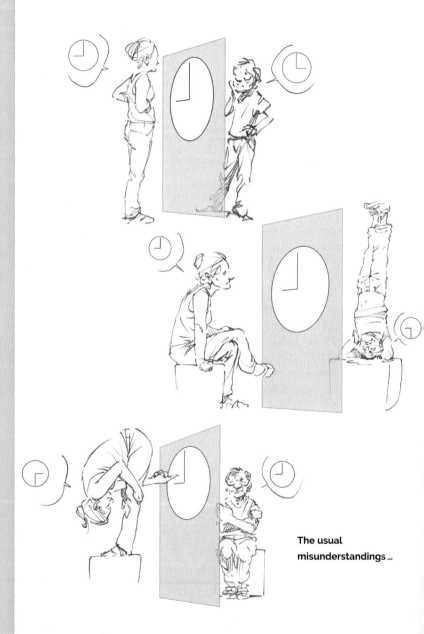

The usual
misunderstandings ...

Clockwise

The sun is turning in a clockwise direction. This much seems clear. After all, there is even a scout's rule outlining how one can find south quite accurately with the help of an analogue wristwatch (back in the day, these used to be the only tools available!): by pointing the hour hand in the direction of the sun and bisecting the angle to the 12-hour position. The reason for this? The hand makes two complete turns around the clock in 24 hours, thus moving at twice the pace of the sun.

However, in the age of globalisation, there is a problem with this approach: with the described rule, it is possible to find *north* in the southern hemisphere. That part is correct – after all, the sun is standing in the north at noon in these regions. In addition, it is moving *counter-clockwise*. The explanation for this is quite simple: we are standing "upside down" in the southern hemisphere, and therefore, everything rotates in the opposite direction.

Does it really, though? Let's make a headstand and look at the second hand of our analogue watch. (If we tracked the minute or the hour hand, we would become red in the face because it would take ages.) The hand is still moving clockwise. There must, then, be another reason.

Well, if we want to observe the sun, we of course intuitively look in the direction of the sun, which is standing in the south in the northern hemisphere and in the north in the southern hemisphere. We are thus looking in the opposite direction!

Objects can only turn around their own axis in space. If we look in the direction of the axis and determine the orbit, it is exactly the reverse compared to looking *away* from the direction of the axis. The sun (being a fixed star) only appears to be turning because we ourselves are rotating around the earth's axis. When we look towards the south, we are essentially looking in the *opposite* direction of the earth's axis; however, when we look towards the north (in the southern hemisphere), we are looking *towards* the earth's axis.

Now let's try the same with the other fixed stars. That shouldn't be a problem: they are rotating clockwise in the northern hemisphere, too. When do we see the rotation of the stars? At dark, of course. Now, however, the sun stands in the north (albeit below the surface of the earth). Now we need to target the North Star. Then, the seemingly complicated motion of the stars becomes quite simple: it is a *counter-clockwise* rotation. In the southern hemisphere, it's the other way round. Now we get it …

With this in mind, we are of course a bit sceptical when a reporter claims that the London Eye is always moving clockwise. Through the eyes of the cameraman, that's certainly true …

Of dwarfs and giants

The drawing shows a situation which we don't think happens that often in our daily life: a miscalculation of size and proportions. Did you chuckle when looking at it? At the dustman or at the alleged dwarf? Let's have a closer look at the situation.

Let us assume that the road indeed narrows down towards the back and that the vociferous person there were a mischievous leprechaun. In that case, we would believe that the road is wider – and in that sense, the dustman in the foreground, who may be aware of this, would be right! However, if the silhouette of the person in the back is of regular height, we would have to assume that the binman has not grasped the principle of perspective.

As humans, we are not equipped with an integrated laser beam that would enable us to measure distances between persons or objects. This is the first thing we should measure if we want to know the size of an object. Next, we should determine the *angle* under which the object appears. Measuring angles works much better than measuring distances when we rely on our eyes only. When measuring distances, we use *both eyes*, which means that the targeted point appears under slightly different horizontal angles and the substantial experience we gain with such measurements over the course of our lives helps resolve this discrepancy.

Let's put this to the test using a scene that we've all experienced multiple times: we are driving on a road beside a vast green area free of buildings or trees, where dozens of wind turbines are generating "sustainable energy", a common buzz word. If you're being honest, you have no idea how big these turbines are (we didn't know it off the cuff either): nowadays, they can reach a height of over 200 metres (including the rotor blades). As exercise, you can now calculate how quickly the ends of the wings are moving if they complete a full rotation in three seconds. Should you have guessed a much smaller height for some reason, just multiply your result with the "factor of error".

Another classic example features a "moon illusion": the moon, of course, does not change its size in the course of the night. What does change is the visual angle under which it appears, which is usually $1/2°$ – that means that you could easily cover the moon with your thumb if you stretch out your arm, even if it were four times as large. If the moon has, however, risen and looks "huge" between two trees that are standing far apart, you will be surprised at seeing the moon in the sky a few hours later, where it will once again appear quite small.

**Everything under control
from all perspectives?**

The mirror paradox multiplied

What happens when we make some final adjustments to our outfit in the mirror? We lift our left hand to the left eye. Our very realistic mirror image instantly repeats this movement, but lifting the right hand to the right eye: are left and right reversed here?

Mathematically speaking, the distances from the mirror become "negative distances". So, my image in the mirror is not entirely identical to me. However, since we tend to believe that it is, we get the impression that left and right have been reversed. This can be quite confusing …

You have no doubt been in an elevator with a cabin that has mirrors on the left and right walls. In such elevators, you can see yourself "infinitely many times". But wait a second – sometimes you see yourself fromt he front, other times from behind, then from the front again, etc. So, you actually see yourself "infinitely many times twice". In the second mirror image, you see yourself from behind, but in the mirror in front of you, you get a front view. Should something crawl up your left shoulder, it will also do so in the second mirror image.

If there is an additional mirror on the third wall of the elevator cabin, the effect is increased: When you look, for instance, at the corner of the left mirror (where the left and the back mirror meet at a right angle), you will – somewhat unexpectedly – see another reflection of yourself, only this time it is not a mirror-inverted reflection but one that has been rotated by 180°. It looks as if your double is trying to follow your verbal instructions very precisely: "Left hand to the left temple" …

In this case, our mirror image is made up of two *double* reflections: The one half that you see to your left comes from your clone at the back wall, which is actually standing to your right but is reflected in the left mirror. The other half on the right is the reflection of your mirror image that is actually standing to your left in the left mirror but has been reflected to the back mirror. Are you still following? The whole thing resembles a spaghetti western, where the armed villain enters a mirror cabinet and tries to blow the hero, who appears multiple times, to kingdom come with a single bullet.

If the ceiling of the elevator cabin is also mirrored, things get even more exciting. If you look, for instance, at the top-left corner, then you will see another reflection of yourself that is not only mirror-inverted but also upside down. What you see this time are three triple reflections that combine into one mirror image. Shall we leave the verbal breakdown of this for some other time?

If you place such a mirrored cube corner on the moon and shoot at this corner with a laser beam from an arbitrary point on earth, then the laser beam will return to you – after being reflected three times – in about two seconds. This method is regularly used to measure the changing distance from the earth to the moon.

Let's see who will catch whom ...

Fish eye and fisheye perspectives

We cannot see under water without some sort of auxiliary device: The refractive index of our cornea is too low to focus the incident light rays in the water onto our retina. Sea lions, on the other hand, are very short-sighted on land, but they see very well under water. Diving googles help us humans do away with this problem. Those who wish to be very cool could splurge on amphibious contact lenses, which have an extremely high optical power and thus allow you to see under water like a crocodile (which have a translucent third eyelid, the so-called nictitating membrane, for this purpose). However, such lenses often produce tight lens syndrome, as they tend to shrink during diving and then fit too tightly on the cornea.

If the the water surface is completely undisturbed, aquatic animals can see everything above the surface compressed into a circle. This circular area is called Snell's window and has been named after the person who discovered the law of refraction, also known as Snell's law. Its peculiar refraction results from the fact that light, which travels in air at practically the same speed as in a vacuum, is $1/4$ slower in water.

If we take a photograph from a shallow water depth and pointing a strong wide-angle lens vertically upward, the picture that we get from the above-water world will look like a photograph taken with a fisheye lens: Things at the centre of the frame will appear fairly normal, but towards the circular margins, the image becomes more contracted. Crocodiles, which often lie in wait in shallow waters, can thus assess precisely when and how they must launch themselves out of the water in order to catch their prey. For their prey, on the other hand, it is difficult to detect the predator because the water surface is highly reflective when seen from a shallow angle.

What do you see outside of Snell's window? While everything around you appears undistorted as usual under water, outside of the circle – and once again, with an undisturbed water surface – you see, beginning almost seamlessly, the total reflection of those objects that are far enough away to be mirrored completely.

Each point underwater has "its own Snell's window", through which it is "supplied" with sunlight. Essentially, only this light can be reflected. This means that if a fish A swims just below the water surface, it will be hit by almost the full spectrum of sunlight. If the fish is watched by a crocodile B from a distance of, say, 10 metres, then the light that is reflected by A must travel 10 metres to B. After 10 meters, the red wavelengths of sunlight will already disappear. So, from B's perspective, A will usually appear in blue and green shades. However, if fish A is red, it will be almost invisible to B. Red is thus the ideal camouflage colour under water, especially at greater water depths.

If, however, the crocodile looks towards the surface from a shallow depth, the world above water will appear bright and colourful to it through Snell's window.

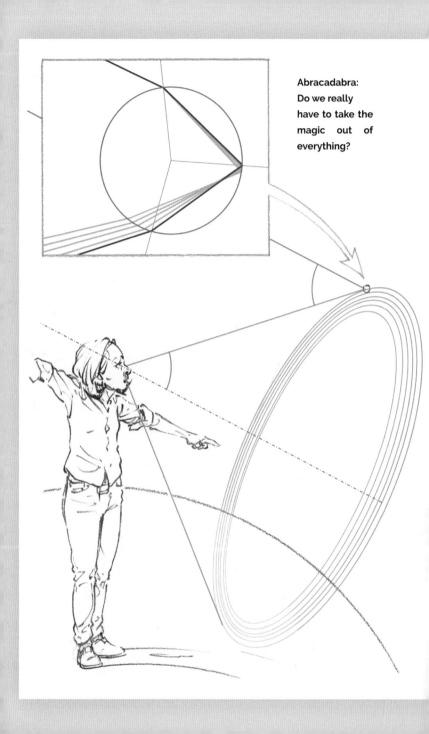

Abracadabra:
Do we really
have to take the
magic out of
everything?

The treasure at the end of the rainbow

Spotting a rainbow is always delightful. Legend has it – and though it is only a legend, it is still a nice one to believe in – that there is a treasure hidden at the place where the rainbow meets the ground: *Abracadabra*!

This magic word was already used by the Roman physician Quintus Serenus. He prescribed wearing amulets with 'abracadabra' inscribed on them, claiming that it would offer protection against malaria. The magic formula still exists today in several languages (even in Russian). Abracadabra's malaria-healing magical powers have since been refuted, but why should we also spoil the more innocuous mystery of rainbows?

In the 17th century, the French philosopher, mathematician, and natural scientist (a classical combination) René Descartes, to whom we owe the now widely established Cartesian coordinate system, already managed to produce an explanation for the appearance of rainbows:

First of all, the basic premise: When travelling from an optically denser medium to a rarer one, a sunray is split into the colour spectrum of a rainbow. Today we know why: sunrays consist of light with varying wave lengths, and each wave length has its own index of refraction.
In the case of the rainbow, the optically denser medium is water. A rain shower on a warm day produces millions of tiny water droplets. Due to surface tension, these droplets have the exact shape of a sphere. Descartes realized that parallel sunrays entering rain drops are refracted inside the droplets and a part of them is then total-reflected off the droplet's back wall. When the sunrays exit the water spheres in the opposite direction, we get the spectral colours of the rainbow, which scatter from the incident sunlight at an angle of $\alpha = 42° \pm 2°$, depending on the colour.

The coloured light of rainbows is thus bounced back by countless millions of raindrops, but we can only see the colours when we look at the droplets from the angle α. At this angle, the droplets seem to form a circle. To be more precise, they lie on a cone with the half-angle of aperture α. The cone axis passes through our eye in the direction of the sunrays. The cone can only partially be seen and disappears on the horizon. When we change our position, the cone moves with us, and so does the end of the rainbow. That is why it is impossible to find the treasure that is supposed to be hidden there.

Question: Does this make rainbows any less fascinating or beautiful? On the contrary: We will be able to enjoy their spectacle more often, because now we know where – that is, at what angle – to find them.

Rolling shutter self-experiment

How real is our reality?

If you are a photographer who is not only interested in taking classical pictures and shooting motionless or lightly moving objects under normal conditions, you might end up taking pictures that give a distorted impression of reality. You might, for instance, stand at the finish line of a horse race, equipped with an outrageously expensive camera and fully focused on capturing the moment the two frontrunners cross the line, and although the picture that you take with an exposure time of $1/10000$ seconds seems to "prove" that horse B was ahead of horse A by a nose, you might still be wrong!

At the finish line, it would probably have been better to use a longer exposure time – your picture might then be blurry, but you would avoid a blunder that could have terrible consequences, such as an expensive lawsuit after submitting an unsuccessful appeal for nullity to the racing commission.

The technical term for this type of photography is "rolling shutter effect". To get a better understanding of this effect, let us do a little self-experiment: you will need a regular photocopier. During the copying process, you can see a horizontal bar moving from left to right within a second and scanning the document that has been placed on the glass pane "line by line". This process can produce a razor-sharp $1 : 1$ reproduction of your letter to the racing commission's lawyer. So far, so good. Now place your left hand on the left-hand side of the photocopier's glass pane and use your right hand to press the copy button. Once the horizontal bar passes under your left hand, turn that hand to the right-hand side of the glass plane and let the horizontal bar scan it a second time. The result will bear some resemblance to the drawing on the left page – though here it was the head of our test subject that was "folded" to the other side. More advanced experimenters could try to turn their hands several times to create more artistically distorted images combining multiple exposures of the real world.

99.999999% of all sensors in photographic cameras capture images in basically the same manner as photocopiers. The image is exposed and captured line by line (rolling shutters move from top to bottom). An advantage of this process is that it allows for an extremely short exposure time (e.g. $1/10000$ seconds). Let us now take another picture from the grandstand of the racing court to capture the two horses crossing the finish line. Horse A is farther away from you and reaches the finish line four centimetres before horse B. The moment horse A crosses the finish line with its head, you press the shutter. It takes the shutter – let's say – one millisecond to reach the head of A. In the meantime, A has run 2 more cm. Until the shutter reaches the head of horse B, five further milliseconds have passed and B has run 10 more cm, so that B falsely appears to have won.

A solution to this problem is offered by the so-called "global shutter", which is, however, so expensive at the moment that barely anyone can afford it …

Index

Georg Glaeser
Head, Institute of Geometry, University of Applied Arts Vienna, Austria
Markus Roskar
Institute of Geometry, University of Applied Arts Vienna, Austria

Library of Congress Control Number: 2019937128

Bibliographic information published by the German National Library The German National Library lists this publication in the Deutsche Nationalbibliografie; detailed bibliographic data are available on the Internet at http://dnb.dnb.de.

© 2019 Walter de Gruyter Gmbh, Berlin/Boston

Translation Tamara Radak & Eugenie Theuer
Copy Editing Tamara Radak & Eugenie Theuer
Graphic Design Peter Calvache

Project Management "Edition Angewandte" on behalf of the University of Applied Arts Vienna: Anja Seipenbusch-Hufschmied, A-Vienna
Content and Production Editor on behalf of the Publisher: Angela Gavran, A-Vienna

Printing: Christian Theiss GmbH, A-9431 St. Stefan

ISSN 1866-248X
ISBN 978-3-11-066354-9
This book is also available in a German language edition (ISBN 978-3-11-066240-5).

www.degruyter.com